SNEAKY
SCIENCE
TRICKS

SNEAKY SCIENCE TRICKS

Perform sneaky mind-over-matter,
levitate your favorite photos, use water to detect
your elevation, navigate with sneaky observation
tricks, and turn a cereal box into a collapsible
robot with everyday things

Cy Tymony

**Andrews McMeel
Publishing, LLC**
Kansas City • Sydney • London

10 11 12 13 14 RR2 10 9 8 7 6 5 4 3 2 1

ISBN-13: 978-0-7407-7398-3
ISBN-10: 0-7407-7398-4

Library of Congress Control Number: 2009943989

www.andrewsmcmeel.com
sneakyuses.com

Certified Chain of Custody

60% Certified Fiber Sourcing and
40% Post-Consumer Recycled

www.sfiprogram.org

The SFI label only applies to the text stock.

ATTENTION: SCHOOLS AND BUSINESSES

Andrews McMeel books are available at quantity discounts with bulk purchase for educational, business, or sales promotional use. For information, please write to: Special Sales Department, Andrews McMeel Publishing, LLC, 1130 Walnut, Kansas City, Missouri 64106.

DISCLAIMER

This book is for the entertainment and edification of its readers. While reasonable care has been exercised with respect to its accuracy, the publisher and the author assume no responsibility for errors or omissions in its content. Nor do we assume liability for any damages resulting from use of the information presented here.

This book contains references to electrical safety that *must* be observed. *Do not use AC power for any projects listed.* Do not place or store magnets near such magnetically sensitive media as videotapes, audiotapes, or computer disks.

Disparities in materials and design methods and the application of the components may cause results to vary from those shown here. The publisher and the author disclaim any liability for injury that may result from the use, proper or improper, of the information contained in this book. We do not guarantee that the information contained herein is complete, safe, or accurate, nor should it be considered a substitute for your good judgment and common sense.

Nothing in this book should be construed or interpreted to infringe on the rights of other persons or to violate criminal statutes. We urge you to obey all laws and respect all rights, including property rights, of others.

CONTENTS

Acknowledgments.. xi

Introduction.. xiii

SNEAKY SCIENCE TRICKS . . . 1

Bernoulli Principle Tricks.. 2

Sneaky Helicopter.. 13

Hand-Powered Fan.. 15

Sneaky Balancing Tricks.. 17

Sneaky Rollback Toy.. 24

Sneaky Switch 27

Detergent-Powered Fish 29

Coin Dancing on Bottle Top 31

Color from Black & White 33

Break Strong String without Scissors....................................... 35

Paper Banger... 37

Talk on a Light Beam 39

Sneaky Walk-Along Glider.. 42

Sneaky Vibra-Bot... 45

Sneaky Vibra-Boat 49

Sneaky Trashformer Robot.. 51

SNEAKY MEASUREMENT PROJECTS . . . 61

Sneaky Weather Barometer.. 62

Sneaky Altimeter.. 65

Make an Anemometer.. 68

Make a Hypsometer.. 71

Sneaky Voltmeter 75

SNEAKY ASTRONOMY AND NAVIGATION TRICKS . . . 79

Make a Sneaky Compass 81

Sneaky Sundials.. 83

Sneaky Quadrant.. 88

Sneaky Direction Finding: Use a Watch................................ 92

Sneaky Direction Finding: Use the Stars 93

Sneaky Direction Finding: Use a Stick 94

SNEAKY MAGIC TRICKS . . . 97

Balancing Soda Can.. 98

Magically Join Two Books... 99

Magical Money Motor.. 101

Static Electricity Tricks... 103

Jumping Tadpole Origami... 106

Sneaky Animated Cassette Tape Creation 108

Sneaky Soda Can Refill 110

Sneaky Money Balance Trick 112

Sneaky Floating Photos 114

Dual Floating Photo Display... 117

Sneaky Magic Wand 120

Sneaky Mind-over-Matter... 122

SNEAKY TRIVIA . . . 129

Water on a Penny.. 130
The Height of the Eiffel Tower 131
Body Measurements 132
Statues 132
Russia and America 133
Alaska... 133
A Boeing 747's Wingspan.. 135
Bamboo.. 135
Watermelons in Japan 136
The Bloodhound.. 136
Mosquitoes 137
The Horned Lizard.. 137
Shark Embryos... 138
Hummingbirds.. 139
The Mayfly... 140
Magnetic Birds 140

World City Latitude List... 141
Popular U.S. City Latitude List.. 144

ACKNOWLEDGMENTS

I'd like to thank my agents, Sheree Bykofsky and Janet Rosen, for believing in my Sneaky Uses book concept from the start. Special thanks to Katie Anderson, my Andrews McMeel editor, for her valuable support. I'm also grateful to the following people who helped spread the word about the first five Sneaky Uses books: Ira Flatow, Gayle Anderson, Susan Casey, Mark Frauenfelder, Sandy Cohen, Katey Schwartz, Cherie Courtade, Mike Suan, John Schatzel, Melissa Gwynne, Steve Cochran, Christopher G. Selfridge, Timothy M. Blangger, Charles Bergquist, Phillip M. Torrone, Paul and Zan Dubin Scott, Dana Vinke, Cynthia Hansen, Charles Powell, Harmonie Tangonan, and Bruce Pasarow. I'm thankful for project evaluation and testing assistance provided by Sybil Smith, Isaac English, and Bill Melzer. And a special thanks to Helen Cooper, Clyde Tymony, George and Zola Wright, Ronald Mitchell, and to my mother, Cloise Shaw, for providing positive motivation, resources, and support for an early foundation in science, and a love of reading.

INTRODUCTION

Sneaky Science Tricks will make your initial entry into science an extraordinary venture. It includes insight into the principles behind its projects, offering you even more sneaky knowledge.

For instance, after you make a Sneaky Boomerang, you'll learn how it uses the Bernoulli principle to stay aloft. This scientific principle not only explains how birds and airplanes can fly but it also permits pitchers to throw sneaky sinking balls, improves automobile stability with rear spoiler, and lets sailboats sail directly into the wind. You'll fascinate your family and friends when you talk on a light beam, display gravity-defying toys, levitate objects, and do other projects that seem like magic.

You'll learn sneaky sources for wire and how to connect things. You'll see how to escape almost any grasp, make clever center-of-gravity balancing designs and sneaky boomerangs, and create a ring that controls devices.

Sneaky Trivia is also included so you can stump your family and friends with little-known geography, history, and nature facts. They will be astonished to learn which animal can shoot blood from its eyes; how fast bamboo grows in a day; how much the height of the Eiffel Tower varies in a day; which single state is the farthest north, south, or east; and how many drops of water can fit on a penny (over thirty!).

All of the projects have tested safe and can be made in no time. If you have an insatiable curiosity for sneaky secrets of everyday things, look no further. You can start your entry into clever resourcefulness here.

SNEAKY SCIENCE TRICKS

Are you stumped trying to come up with a different kind of science-project idea? Or do you want to repurpose some throw-aways into useful educational toys? This section presents over a dozen science tricks that will help you learn and demonstrate sneaky science principles easily. Virtually all of the items needed are already in your home, and the projects can be put together quickly.

Discover Bernoulli's principle while you make flying boomerangs and Frisbee-like disks. Store energy in a rubber band and make a rolling can magically reverse direction and roll back to you, even uphill. Use a magnet to mysteriously activate lights, buzzers, or toys.

Ever think you could turn a black-and-white picture into color? Or talk on a light beam or break tough string without scissors? Well, you can. You'll see how to make a walk-along glider that floats in front of you and how to turn cereal boxes into a mobile robot.

All of these sneaky science techniques are right here ready for quick assembly. Let's get started!

BERNOULLI PRINCIPLE TRICKS

Sneaky Demonstrations of Air Pressure and Wing Lift

Have you ever wondered how airplanes and helicopters are able to fly? If you have, and want to demonstrate this principle, all you need are such ordinary items as straws, postcards, and strips of paper.

Air Pressure Demonstration I

An ordinary straw can be used to demonstrate that air pressure is all around us (15 pounds per square inch, to be exact). You can demonstrate this easily enough with everyday items.

What's Needed

▶ Straw
▶ Glass filled with water

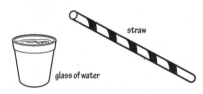

straw

glass of water

What to Do

Insert a straw into the glass of water, as shown in **Figure 1**. Next, place a finger over the top of the straw and lift it out of the water. See **Figure 2**.

You'll see that the water stays in the straw and doesn't flow out because air pressure from the bottom is keeping it in, as shown in **Figure 3**. When you lift your finger from the top of the straw, air pressure flows from the top and pushes against the water, forcing it out.

FIGURE 1

Place straw in water.

FIGURE 2

Hold top of straw with finger.

FIGURE 3

Water stays in straw because of air pressure at bottom.

Air Pressure Demonstration II

You can demonstrate the power of air pressure in a more dramatic way with the following project, again using everyday items.

What's Needed

▶ Glass filled to the brim with water
▶ Plastic-coated postcard

glass of water

postcard

What to Do

Working over a sink, hold up the glass of water. Place a postcard over the mouth of the glass and turn the glass upside down, holding the postcard in place with your finger under it, as shown in **Figure 1**.

Carefully remove your finger from the postcard and you should see that the postcard will not fall. With no air in the glass to push against the postcard, the air outside presses against the postcard, keeping it in place, even with the weight of the water upon it. See **Figure 2**.

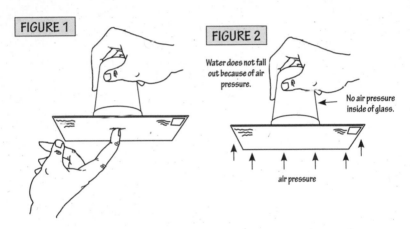

FIGURE 1

FIGURE 2

Water does not fall out because of air pressure.

No air pressure inside of glass.

air pressure

Air Pressure Demonstration III

What's Needed

▶ Paper (preferably a paper towel or napkin)
▶ Scissors

What to Do

Cut a paper strip 1/2 inch wide by 4 inches in length as shown in **Figure 1**. Hold the paper strip up to your face above your mouth and blow. The paper naturally moves upward. Now hold the paper strip just below your lips and blow above the strip. As shown in **Figure 2**, the paper will also rise and move upward!

This occurs because of Bernoulli's principle, which states that fast-moving air has less pressure than nonmoving air. The air under the strip has more pressure than the air above it and pushes the strip upward.

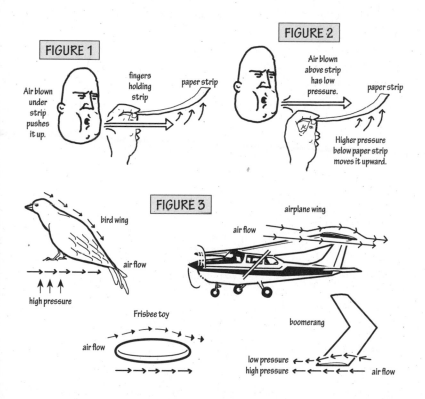

Figure 3 illustrates a side view of a bird's wing, an airplane wing, a Frisbee flying disk, and a boomerang. Notice the top of the wing curves upward and has a longer surface as compared to the bottom. When the airplane moves forward, air moves above and below the wing. The air moving along the curved top must travel farther and faster than the air moving past the flat bottom surface. The faster-moving air has less pressure than the air at the bottom and this provides lift.

Baseball pitchers can take advantage of Bernoulli's principle by releasing the ball with a forward spin. The ball produces a lower pressure below it, causing it to dip when it reaches the plate. Hence, a curveball. See **Figure 4**.

Sailboats apply Bernoulli's principle to use the wind, regardless of its direction, to propel the boat in any desired direction. **Figure 5** shows how altering the shape of the sail into a curve produces an effect similar to that of an airplane wing. The wind moves at a faster rate over the curved side, with a lower pressure, and the higher pressure on the other side of the sail pushes the boat laterally. A centerboard, attached to the boat hull, prevents the boat from moving sideways while allowing it to use the wind thrust to move forward. See **Figure 6**.

Automobile bodies are similar to an airplane wing because they are flat on the bottom and curved on top. They can lose stability at high speeds since they tend to achieve lift from the higher air pressure below, as shown in **Figure 7**. To reduce the Bernoulli effect, automakers have incorporated improvements in vehicle design, such as lowering the body height, adding special front bumper and fender contours, and installing rear spoilers. See **Figure 8**.

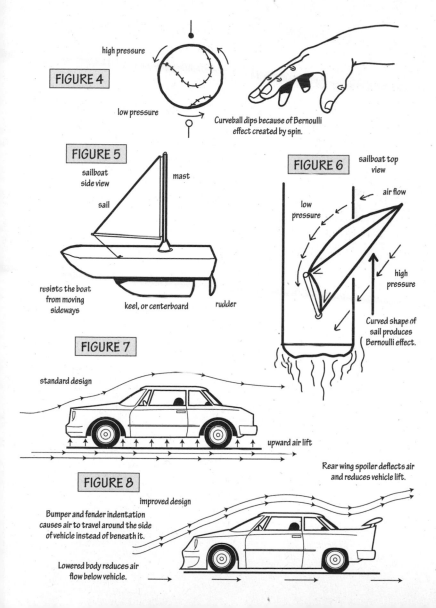

FIGURE 4

high pressure

low pressure

Curveball dips because of Bernoulli effect created by spin.

FIGURE 5

sailboat side view

mast

sail

resists the boat from moving sideways

keel, or centerboard

rudder

FIGURE 6

sailboat top view

air flow

low pressure

high pressure

Curved shape of sail produces Bernoulli effect.

FIGURE 7

standard design

upward air lift

Rear wing spoiler deflects air and reduces vehicle lift.

FIGURE 8

improved design

Bumper and fender indentation causes air to travel around the side of vehicle instead of beneath it.

Lowered body reduces air flow below vehicle.

Air Pressure Demonstration IV

What's Needed

- Scissors
- Paper (preferably a paper towel or napkin)
- Two empty soda cans
- Magazine

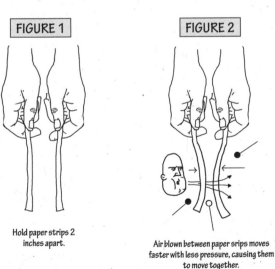

scissors

paper towel

magazine

soda cans

What to Do

Cut two paper strips 1/2 inch wide by 4 inches in length and hold them about 2 inches apart, as shown in **Figure 1.** Blow air between the paper strips and watch what occurs. You would expect the strips to blow apart but they actually come together, as shown in **Figure 2.**

FIGURE 1

Hold paper strips 2 inches apart.

FIGURE 2

Air blown between paper srips moves faster with less pressure, causing them to move together.

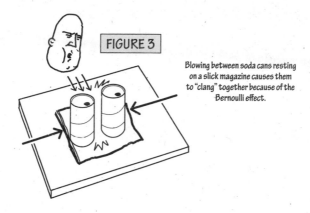

Blowing between soda cans resting on a slick magazine causes them to "clang" together because of the Bernoulli effect.

Bernoulli's principle is working here because the faster-moving air blown between the paper strips has less pressure than the air on the other side of the paper. This higher pressure pushes the strips toward each other.

Now, place the two empty soda cans an inch apart upon the slick surface of a magazine. When you blow between the cans, they will move toward each other, producing a clanging sound. See **Figure 3**.

Air Pressure Demonstration V

Here's another sneaky, easy-to-perform demonstration of air pressure's causing an unexpected result.

What's Needed

▶ Scissors
▶ Piece of paper

scissors   paper

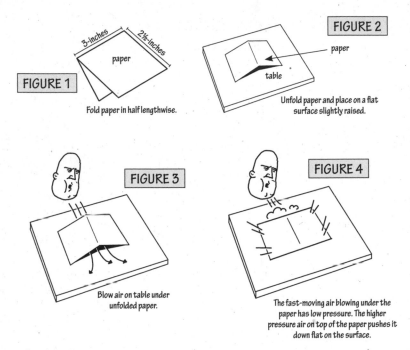

FIGURE 1
Fold paper in half lengthwise.

FIGURE 2
paper
table
Unfold paper and place on a flat surface slightly raised.

FIGURE 3
Blow air on table under unfolded paper.

FIGURE 4
The fast-moving air blowing under the paper has low pressure. The higher pressure air on top of the paper pushes it down flat on the surface.

What to Do

Cut the piece of paper into a 5 by 3-inch shape. Fold the paper in half lengthwise, as shown in **Figure 1**.

Next, unfold the paper and place it on a flat surface so that it has a slight rise near its center crease. See **Figure 2**.

Then, as shown in **Figure 3**, bring your face close to the surface of the table and blow underneath the paper.

You would expect the paper to rise but it actually flattens downward. The higher air pressure on top of the paper, compared to the fast-moving air beneath it, pushes the paper flat on the table, as shown in **Figure 4**.

Air Pressure Demonstration VI

You can use Bernoulli's principle to perform a neat magic trick by making a ball rise from a cup and jump into another one without touching it.

What's Needed

Ping-Pong ball

small cups

▶ Ping-Pong ball
▶ Two small cups

What to Do

This project requires small cups that are slightly smaller in diameter than the Ping-Pong ball. Since the Ping-Pong ball can barely fit in the cup, rapidly moving air above the ball will not affect the air pressure beneath it.

Put the ball into one of the cups and place it about 3 inches away from the second cup, as shown in **Figure 1**. Blow as hard as you can above the first cup and the ball should start to rise. See **Figure 2**. The force of your breath will push the raised Ping-Pong ball over to the empty cup, where It will drop inside, as shown in **Figure 3**. With a little practice, you can make this sneaky trick work every time.

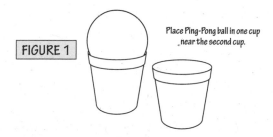

FIGURE 1

Place Ping-Pong ball in one cup near the second cup.

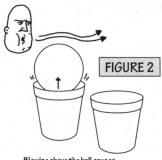

Blowing above the ball causes
it to rise.

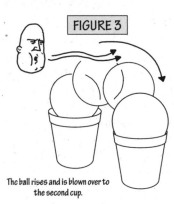

The ball rises and is blown over to
the second cup.

SNEAKY HELICOPTER

Unlike airplanes or gliders, helicopters use a rotating set of wings to achieve lift and can float safely to the ground without engine power. You can demonstrate this with just a single piece of paper and make a simple toy.

What's Needed
▶ Paper
▶ Ruler
▶ Scissors

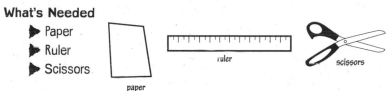

paper   ruler   scissors

What to Do
Cut a piece of paper into a 1 by 7-inch rectangle. Cut slits on opposite sides, 2 inches from the top of the sneaky copter, as shown in **Figure 1**. Fold the paper strip in half and slide the slits together to secure it. Next, bend back the top of the paper above the slits to form blades for the sneaky copter, as shown in **Figure 2**.

Now you can toss the sneaky copter straight up in the air and it will slowly spin downward, because the blades slow it down as they turn. See **Figure 3**. Experiment with the shape of the blades by pushing them farther back or angling them to achieve various downward flight patterns.

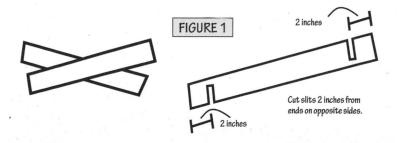

FIGURE 1

2 inches

Cut slits 2 inches from ends on opposite sides.

2 inches

FIGURE 2

Bend strip into a
loop and slide slits
into each other.

FIGURE 3

Drop helicopter from
a tall height and
watch it "Auto-Gyro"
to the ground.

HAND-POWERED FAN

As you may know, hot air rises. Rising heat can be made to move objects, and you can demonstrate this fact with a novel "hand-powered" motor. In this demonstrational science project, your hands will actually provide the heat to show how moving air currents can move an object in a rotary motion.

All it takes is an ordinary piece of paper, scissors, a needle, a cardboard box, and your hands.

What's Needed
- ▶ Paper
- ▶ Scissors
- ▶ Sewing needle
- ▶ Small cardboard box

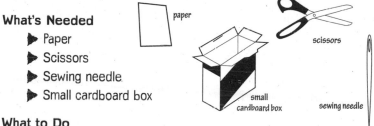

What to Do

Cut a piece of paper into a 2-inch square. Fold it in half diagonally; then unfold it and fold it in half on the other diagonal, as shown in **Figure 1**. This should create a cross-fold with a center point.

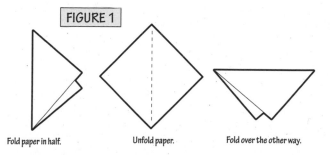

FIGURE 1

Fold paper in half.   Unfold paper.   Fold over the other way.

You can use a paper-clip box or similar small box as a mount for the needle. Hold the needle on its side with your fingers and carefully twist it into the top of the box (or use a thimble) until it punctures a hole in the top. Place the piece of paper on top of the needle so its center point allows the paper to turn freely. See **Figure 2**.

To make the sneaky "motor" turn, rub your hands together back and forth about twenty times to generate heat and place them near the sides of the paper. After a few seconds, the paper will begin to spin (**Figure 3**).

The paper spins because the heat on your hands causes a temperature increase in the air around the paper. As the heated air rises and cooler air takes its place, the air movement pushes the paper sides, causing it to rotate like a motor.

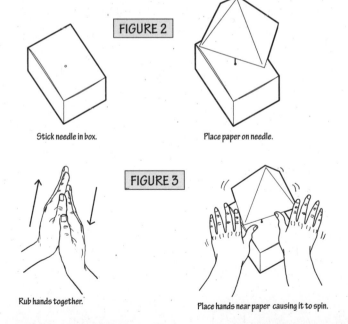

FIGURE 2

Stick needle in box.

Place paper on needle.

FIGURE 3

Rub hands together.

Place hands near paper causing it to spin.

SNEAKY BALANCING TRICKS

You can make everyday things balance in sneaky ways when you know the secret to determining the center of gravity. The center of gravity is the point in an object at which its mass is in equilibrium. Where this point is depends on the object's shape and weight distribution, and you can produce some attention-getting creations with this knowledge.

The following four projects are easy to do with items found just about everywhere.

Sneaky Balancer I

Knowing how to lower the center of gravity of an object allows you to produce figures that seemingly defy gravity (or make you seem like a skilled magician). This project demonstrates what happens when two similar cardboard figures have their center of gravity in different positions.

What's Needed

▶ Scissors
▶ Cardboard, a piece 8 1/2 by 11 inches
▶ Sewing thread (optional)

sewing thread

scissors

cardboard

What to Do

Cut out the small shape shown in **Figure 1** from the piece of cardboard. Follow the dimensions shown. Next, try to balance the head of the figure on your fingertip, as shown in **Figure 2**. It's almost impossible to keep it upright without its tipping over.

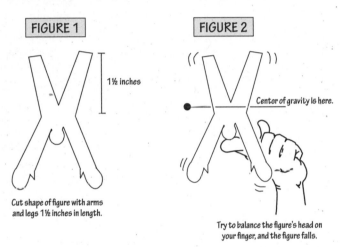

FIGURE 1

1½ inches

Cut shape of figure with arms
and legs 1½ inches in length.

FIGURE 2

Center of gravity is here.

Try to balance the figure's head on
your finger, and the figure falls.

Next, cut out the figure shown in **Figure 3**. The only difference is the legs are much longer. Try to balance this larger figure on your hand. It's easy now, because the center of gravity is below your finger. See **Figure 4**. You should be able to walk around the room and the figure will not fall.

GOING FURTHER:

Acrobats keep their balance using this principle. To demonstrate, cut a small slit in the head of the figure. See **Figure 5**.

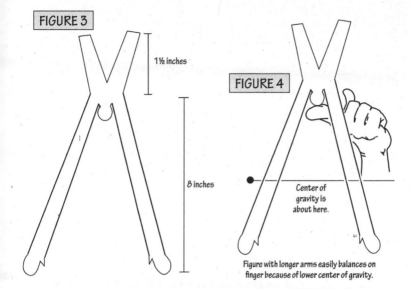

FIGURE 3

1½ inches

8 inches

FIGURE 4

Center of
gravity is
about here.

Figure with longer arms easily balances on
finger because of lower center of gravity.

Then, tie a length of thread from a chair to a lower object, such
as another chair or table, and set the figure on the thread. The figure
should rest on the thread in its slit and, with a slight push, slide across
without falling. See **Figure 6**.

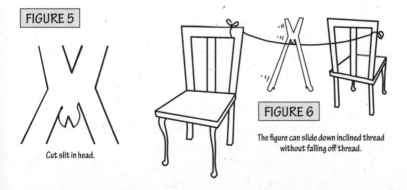

FIGURE 5

Cut slit in head.

FIGURE 6

The figure can slide down inclined thread
without falling off thread.

Sneaky Balancer II

This sneaky balancer can rest horizontally on the tip of a paper clip and will surely astonish onlookers.

What's Needed
▶ Scissors
▶ Cardboard
▶ Paper clip

scissors

cardboard

large paper clip

What to Do
Cut out the figure shown in **Figure 1** from the piece of cardboard. Be sure to include the spiked hair, with a long center spike. Try to adhere to the dimensions shown but, if desired, you can produce a larger or smaller figure as long as you keep the arm and body lengths in proportion.

Bend the figure's arms down at the shoulder and elbows. See **Figure 2**.

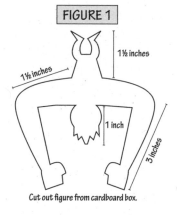

FIGURE 1

1½ inches

1½ inches

1 inch

3 inches

Cut out figure from cardboard box.

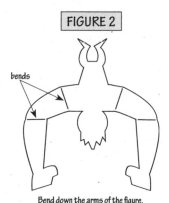

FIGURE 2

bends

Bend down the arms of the figure.

Next, bend a paper clip, as shown in **Figure 3**, so one end stands up vertically.

Last, place the figure on the paper clip with the spiked hair resting on the tip. If necessary, bend the arms down so it won't fall. The figure should balance on the tip of the paper clip. You should be able to carefully push its legs to the right or left and it will stay aloft. See **Figure 4**.

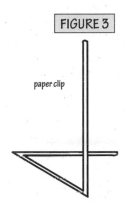

FIGURE 3

paper clip

Send paper clip into a stand for the figure.

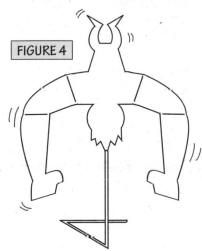

FIGURE 4

Gently rest head of figure on tip of
paper clip and it will magically balance.

Sneaky Balancer III

What's Needed

▶ Scissors
▶ Cardboard, 8 1/2 by 11 inches

scissors

cardboard

What to Do

Cut out the shape shown in **Figure 1** from the cardboard. Be careful to follow the dimensions shown.

You should be able to easily balance the figure on the tip of your finger, elbow, or nose, because its center of gravity is at the large circular area. See **Figure 2**.

You can create similar figures and, using paper clips or coins secured with tape, add weight to an area near the bottom section of the figure so it balances effortlessly.

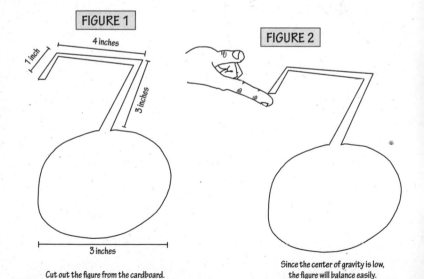

FIGURE 1

4 inches

1 inch

3 inches

3 inches

Cut out the figure from the cardboard.

FIGURE 2

Since the center of gravity is low, the figure will balance easily.

Sneaky Balancer IV

What's Needed
- One quarter
- Two metal forks
- Drinking cup

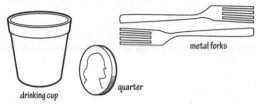

drinking cup   quarter   metal forks

What to Do

Place the quarter between the teeth of the two forks as shown in **Figure 1**.

If a lightweight cup is used, you must fill it with water so it will not tip over. If a heavy cup or jar is used, water is not required.

Carefully rest the edge of the quarter on the lip of the cup. You should be able to let go and the forks will stay aloft. If they don't, adjust the angle of the forks until they balance. See **Figure 2**.

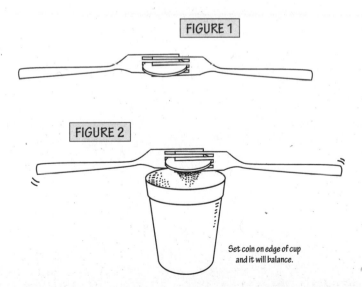

FIGURE 1

FIGURE 2

Set coin on edge of cup
and it will balance.

SNEAKY ROLLBACK TOY

If you've wondered how hybrid cars can boast such impressive miles-per-gallon ratings, this project will show you how. You will demonstrate the principle that allows hybrid cars to store and release energy by using a simple-to-build rollback toy.

What's Needed
▶ Small container
with plastic lid
▶ Extra plastic lid
▶ Thick rubber band
3–4 inches long
▶ Two paper clips
▶ Bolts or large
lug nuts
▶ Scissors

container

plastic lid

rubber band

scissors

paper clips

bolts or nuts

What to Do
First, obtain a small cardboard container and remove the bottom. Cut slits through the center of the can's plastic lid and the extra lid. Take the lids off the container. See **Figure 1**. Thread a rubber band through the bottom of the container and pull it through the lidless top of the container.

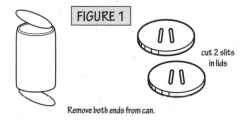

FIGURE 1

cut 2 slits
in lids

Remove both ends from can.

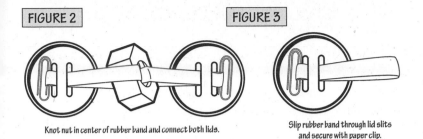

Knot nut in center of rubber band and connect both lids.

Slip rubber band through lid slits
and secure with paper clip.

Tape the washers or lug nuts together and connect them to the
middle of one section of the rubber band (do not tape the strands of
the rubber band together). See **Figure 2**.

Put the end of the rubber band through the container lid. Use a
paper clip to secure the band so it does not slip inside the container
(put the paper clip through the end loop of the rubber band that
remains outside the hole). See **Figure 3**. Put the lid on the container,
making sure the rubber band is still sticking out the other end.

Carefully pull on the rubber band until it comes through the hole.
Secure the band with the second paper clip. Be sure to situate the
weight so it is in the center of the container and does not touch
the sides. Put both lids on the ends of the container. See **Figure 4**.
Your rollback toy is ready to go!

FIGURE 4

Bend and push
one lid through can and
put lids on both ends.

Roll the toy and watch as it stops and returns to you. See **Figure 5**. This sight is more amazing when you roll it downhill and it stops and returns to you uphill, seemingly defying gravity.

The weight holds one end of the rubber band stationary while the free side twists around. The farther the toy rolls, the more potential energy is stored. Release and watch the toy roll back toward you, demonstrating its conversion into kinetic energy.

Hybrid vehicles use this principle to store energy in a flywheel to power an electric motor. The motor is used when you take off from a standing stop to save engine fuel.

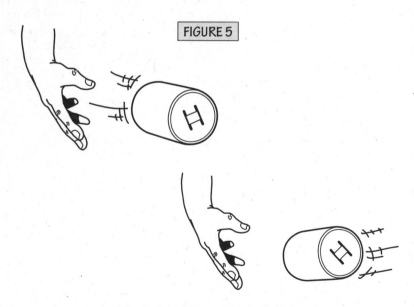

FIGURE 5

Rolling away from you will store energy and it will roll back.

SNEAKY SWITCH

With a simple paper clip and magnet, you can create your very own light switch.

What's Needed

- AA-size battery
- Small 1 1/2-volt lightbulb
- Transparent tape
- Cardboard
- Paper clip
- Stiff copper wire
- Strong magnet

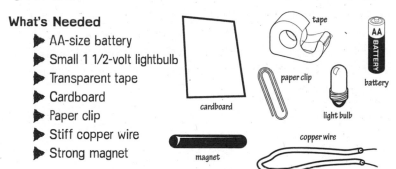

tape

cardboard

paper clip

battery

light bulb

magnet

copper wire

What to Do

In this project, the parts are mounted with tape to a piece of cardboard (a postcard will work fine) so the operation of the Sneaky Switch can be seen by others or recorded on a digital camera.

First, place the battery and bulb end to end and tape them to the cardboard, as shown in **Figure 1**.

Next, bend the paper clip in the shape shown in **Figure 2**, so it wraps around the battery's positive (+) terminal.

Then, wrap the stiff copper wire around the bulb's base and bend it so it hovers over the paper clip, as shown in **Figure 3**. Place a piece of tape on the end of the paper clip to secure it to the board.

Last, ensure that the paper clip is bent upward slightly and is very close but not touching the copper wire above it. It should be able to move freely. Bring the strong magnet close to the top of the copper wire. The paper clip should move upward toward the magnet, contacting the copper wire and lighting the bulb as shown in **Figure 4**.

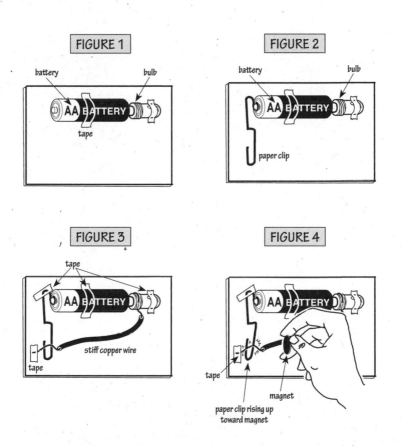

DETERGENT-POWERED FISH

When you think of cardboard and detergent, you don't usually imagine that they could be turned into a self-propelled model fish. This project shows how you can make a fun aquatic toy with household items in just minutes.

What's Needed

small cardboard box

detergent

- Cardboard from a postcard or food container such as a cereal box
- Scissors
- Sink or bathtub filled with water
- Detergent

scissors

What to Do

Cut the shape of a fish from a piece of thin cardboard exactly as shown in **Figure 1**. Note the hole in the center and be sure to include the slit in the tail section.

FIGURE 1

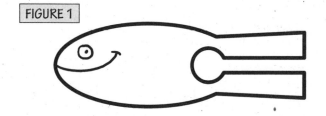

Lay the cardboard fish on the surface of water. Then, carefully pour some detergent onto the circle of the fish. You'll soon see the little fish propel along like a motorboat. See **Figure 2**.

The detergent breaks up the water molecules, which are held together very tightly. As the water loses cohesiveness, it propels the fish model forward, as shown in **Figure 3**.

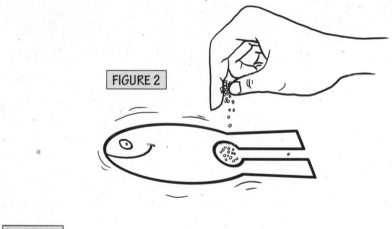

FIGURE 2

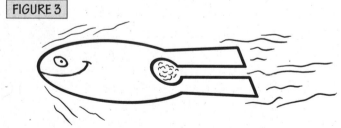

FIGURE 3

COIN DANCING ON BOTTLE TOP

When water's temperature rises, it produces vapor that rises. By placing an object such as a coin in its path, you can make it seem to tap-dance on its own with a little sneaky science. You just need a two-liter bottle, a dime, and your household resources.

What's Needed

▶ Measuring cup
▶ Water
▶ Two-liter bottle
▶ Freezer
▶ Dime

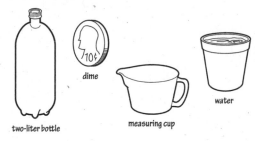

dime

two-liter bottle

measuring cup

water

What to Do

First, pour 1/2 cup of water in the bottle and place it in the freezer for an hour. See **Figure 1**.

Next, remove the bottle and set it upright with the cap removed. Set the dime on top of the bottle so it completely covers the opening, as shown in **Figure 2**.

In a few minutes you'll notice that the dime will start vibrating and making a tapping sound. As the water inside warms, the vapor in the bottle rises and pushes on the dime on the way out. After the vapor escapes, the dime rests on the bottle again. The cycle repeats, which causes the tap-dancing movement, as seen in **Figure 3**.

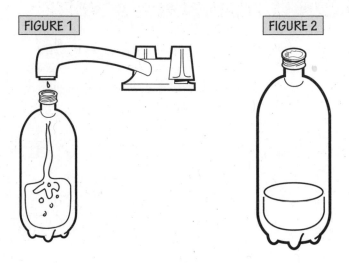

FIGURE 1

FIGURE 2

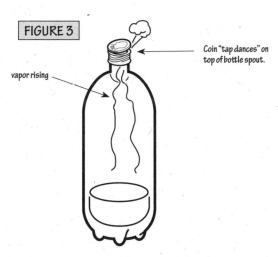

FIGURE 3

vapor rising

Coin "tap dances" on
top of bottle spout.

COLOR FROM BLACK & WHITE

Challenge your friends by asking them to change the color of an item without touching or painting it. Then amaze your friends when you actually produce colors from a black-and-white image with this next trick.

What's Needed

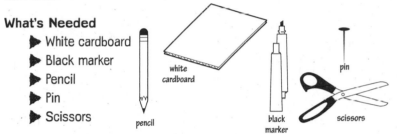

- White cardboard
- Black marker
- Pencil
- Pin
- Scissors

pencil

white cardboard

black marker

pin

scissors

What to Do

Draw the disc shown in **Figure 1** on the white cardboard. Ensure that half the disc is solid black and half has the broken circle picture. If you photocopy the illustration, be sure to fill in blank spots with a marker. The disc should be approximately 4 inches in diameter. Cut out the disc with the scissors.

Place the disc on the center of the pencil eraser. Secure the disc to the eraser by pushing the pin into the eraser through the cardboard. See **Figure 2**.

Next, place the pencil between your palms and spin it. You'll see the black-and-white image turn blue and red depending on the speed, as shown in **Figure 3**.

HOW IT WORKS

Your eyes have rod cells for peripheral, or side vision, and cone cells for front vision and for discerning color.

There are three different types of cone cells that each respond to red, green, and blue at different response times. When you see white, all three cone cells respond equally. When you spin the disc, the alternating black sections, with their different lengths, cause an imbalance to how the three cone cells respond to white and your cone cells cause you to "see" different colors.

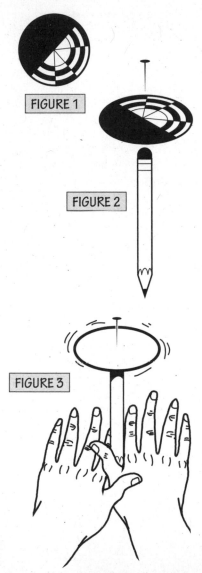

FIGURE 1

FIGURE 2

FIGURE 3

BREAK STRONG STRING WITHOUT SCISSORS

Have you ever failed to break a length of string no matter how hard you tried? Want to look like a hero with super strength when others fail this test? This neat trick will also come in handy when scissors aren't around.

Note: This sneaky trick will work on thin cotton string, not with nylon or coated string.

What's Needed

▶ Cotton string,
  approximately 3-feet long

string

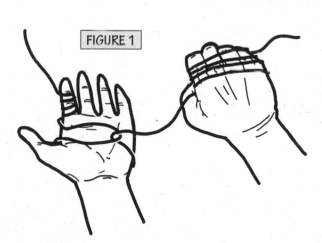

FIGURE 1

What to Do

Wrap one end of the string three times around the forefinger of your left hand and then make a small loop near the palm.

Next, pull the string behind your left hand and up and through the loop, and leave a 1-foot length between your hands.

Last, wrap the string at least four times around your right hand. To break the string, make a strong fist with both hands. See **Figure 1**. Bring your fists together with your left hand on top, then rapidly drop your right hand. The string should break in the area of the loop.

Note: First try this sneaky trick with thin string and then gradually try stronger string.

PAPER BANGER

This popular project has been around for ages. By folding a piece of paper in a particular pattern, you can capture enough air to cause an extremely loud popping sound.

Just follow the simple steps below and you'll be able to create this paper banger anytime.

What's Needed

▶ Letter-sized (8 1/2 by 11-inch) sheet of paper

paper

What to Do

First, fold the left and right sides of the paper toward the center and then unfold them, as shown in **Figure 1**. Next, fold all four corners toward the centerline. See **Figure 2**.

Then, fold the paper toward you, as shown in **Figure 3**, and ensure that it encloses the flaps. Fold down the banger's right side to the bottom. See **Figure 4**. Fold down the left side to the bottom, as shown in **Figure 5**.

Bend the paper back until the two points face toward you and the banger appears to have a triangular shape. See **Figure 6**. Now hold the top two corners together with your fingers and swing the banger down quickly. This will cause air to gather and compress inside the pocket to create a loud bang. See **Figure 7**.

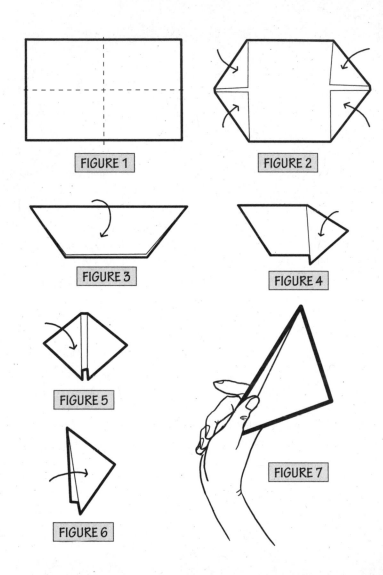

FIGURE 1

FIGURE 2

FIGURE 3

FIGURE 4

FIGURE 5

FIGURE 6

FIGURE 7

TALK ON A LIGHT BEAM

Energy can be converted from one form to another and then detected with sensors. Our ears are audio sensors that detect sounds mostly as vibrations through the air (although you can hear underwater, and by pressing your ear against solid objects, too).

You can make a unique, energy-conversion science project, using a sneaky sound-to-light detector. By vibrating another item, in this case a small mirror, you will be able to detect sounds through a soundproof barrier. Your voice will vibrate a balloon with a small mirror attached. The reflected light vibrations from the mirror are detected by a nearby solar cell, which converts them into electrical signals that can be heard with headphones.

With this project you'll see how sound vibrations can affect light waves and then convert back into sound.

What's Needed

- Solar cell
- Headphones
- Electrical tape
- Scissors
- Wide-mouthed paper cup
- Large, round balloon
- Rubber band
- Small mirror
- Flashlight

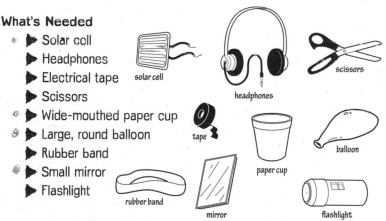

solar cell

headphones

scissors

tape

paper cup

balloon

rubber band

mirror

flashlight

What to Do

Attach the two leads from the solar cell to the end tip and to the upper shaft of the headphone plug and secure the connections with electrical tape. See **Figure 1**.

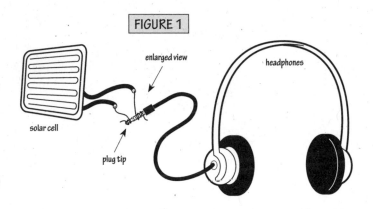

FIGURE 1

enlarged view

headphones

solar cell

plug tip

Next, cut a small hole in the bottom of the cup. Cut the balloon in half, wrap it over the mouth of the cup, and secure it tightly with the rubber band, as shown in **Figure 2**.

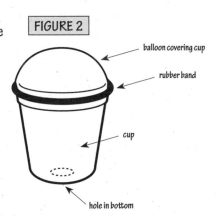

FIGURE 2

balloon covering cup

rubber band

cup

hole in bottom

Tape the small mirror to the center of the balloon. See **Figure 3**. Aim the flashlight at the mirror on the cup and position the solar cell directly in front of the cup, so it receives the reflected light.

Last, gently hold the side of the cup and talk while your friend listens to your voice in the headphones, as shown in **Figure 4**. Test this project while separated by a glass window.

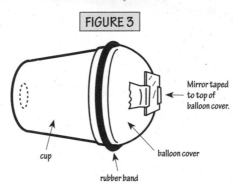

FIGURE 3

Mirror taped to top of balloon cover.

cup

balloon cover

rubber band

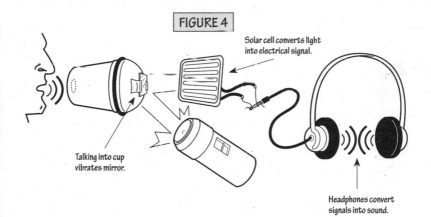

FIGURE 4

Solar cell converts light into electrical signal.

Talking into cup vibrates mirror.

Headphones convert signals into sound.

SNEAKY WALK-ALONG GLIDER

When you walk forward, the air pushes against you; thus you create your own wind. By holding a large piece of cardboard in front of you at just the right angle, you can redirect this wind to provide enough lift for a small, spinning paper glider to float magically in the air.

The sneaky effect is amazing to see and gives the appearance that the glider stays afloat by magic!

What's Needed

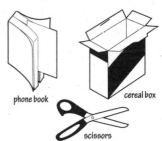

- ▶ Telephone book or tissue paper
- ▶ Scissors
- ▶ Large piece of cardboard, at least 3 square feet, from a cereal box

phone book

cereal box

scissors

What to Do

Note: This sneaky glider must be made out of very thin paper to stay aloft. Ordinary writing paper or newsprint is not light enough.

Cut a piece of paper from a telephone book (use an advertisement page) to the following dimensions: 2 by 4 inches long. Fold up both ends so that 1/2-inch wings stand vertically on each end, as shown in **Figure 1**.

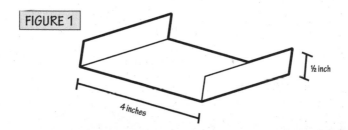

FIGURE 1

½ inch

4 inches

Bend a 1/4-inch section of the middle edge downward, but don't fold all of the way to the end. In the same way, press the other side upward by 1/4 inch. The glider should resemble the picture shown in **Figure 2**.

Unfold the box so it is one wide, flat sheet. Then, hold the glider above your head in a slightly downward, straight position and hold the large cardboard piece with your other hand. See **Figure 3**.

Release the glider and it should roll over downward. Immediately hold the cardboard in a vertical position just behind the paper glider and walk forward, as shown in **Figure 4**. Try to keep the cardboard as straight as possible because the air that reflects from it is what keeps the glider in the air. Tilting the cardboard drastically will cause the glider to fall.

You can steer the glider to the left and right with subtle motions. With some practice you will be able to make the glider stay aloft and guide it along any path you desire.

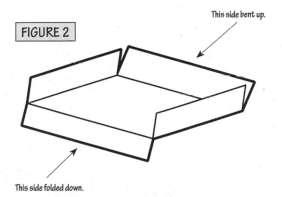

FIGURE 2

This side bent up.

This side folded down.

FIGURE 3

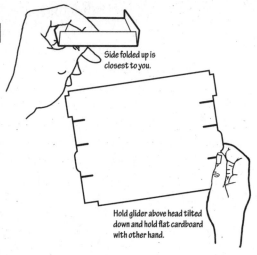

Side folded up is closest to you.

Hold glider above head tilted down and hold flat cardboard with other hand.

FIGURE 4

As you walk forward the glider moves forward and magically floats above cardboard.

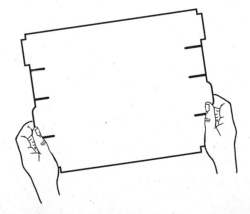

SNEAKY VIBRA-BOT

Normally we want things, such as a motor, to operate smoothly and quietly. But sometimes vibrating objects can serve a useful purpose. Earphones and speakers produce vibrations that you hear as sounds. Vibrating cell phones and pagers alert you by using a tiny motor with an off-center weight attached to its shaft, which causes shaking.

With a small motor removed from an old toy, plus a few other parts, you can make a simple little autonomous robot. This sneaky toy will scamper along any flat surface, without wheels!

What's Needed

- Small toy car or other AA battery–powered vehicle
- Scissors
- Three large paper clips
- Pliers
- Electrical tape
- Styrofoam or paper cup
- AA battery

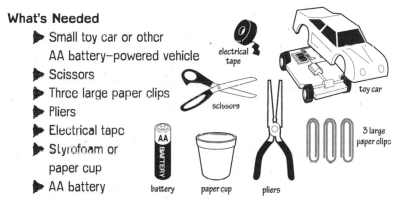

electrical tape

scissors

battery

paper cup

pliers

toy car

3 large paper clips

What to Do

Carefully remove the small motor from the toy car. Be sure to cut away the motor's connecting wires that attach to the battery compartment. See **Figure 1**.

Bend one of the paper clips around the motor's gear into the shape shown in **Figure 2**. With the pliers, tightly press the paper clip securely around the gear. The paper clip's long outstretched end acts as an off-center weight, causing the motor to shake and vibrate as it spins.

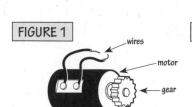

wires

motor

gear

Motor with connecting wires removed from toy.

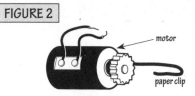

motor

paper clip

Press paper clip around gear with one end hanging off.

Next, bend the other two paper clips into curved leg-stand shapes, as shown in **Figure 3A**. Next, tape them to the sides of the upside-down cup, as shown in **Figure 3B**. They will act as the robot's support legs when it's placed upside down.

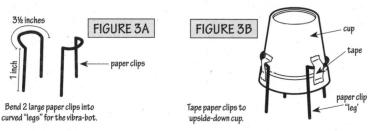

3½ inches

1 inch

FIGURE 3A

paper clips

Bend 2 large paper clips into curved "legs" for the vibra-bot.

FIGURE 3B

cup

tape

paper clip "leg"

Tape paper clips to upside-down cup.

Place the motor on top of the cup so it hangs over the edge. Ensure that the paper clip can turn freely without touching the cup. Tape the motor and battery in place on top of the cup. Then, tape one of the motor's connecting wires securely to one end of the battery. Finally, tape the other wire on top of the battery near the other end. See **Figure 4**.

To activate the robot, press the loose wire against the battery and apply a small piece of electrical tape. This will act as an On/Off switch. When the motor spins, it will vibrate the entire cup and scamper around, as shown in **Figure 5**. To stop the robot, simply pull the wire away from the battery.

FIGURE 4

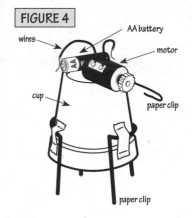

wires
AA battery
motor
cup
paper clip
paper clip

FIGURE 5

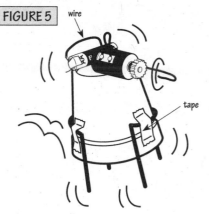

wire
tape

Tape wire to end of battery and cup. Shake and scatter away.

GOING FURTHER:

- Decorate the mini-robot cup with more paper clips, spare plastic, and other odds and ends.
- Instead of a cup, a small box, such as a tea or popcorn box, can be substituted.
- Mount one paper clip slightly lower than the other so the Vibra-bot sprints in a circle pattern.
- Make several Vibra-bots and watch them duel together.

tape

pliers

tea & popcorn boxes

battery

paper cup

cardboard & paper from product packaging

paper clips

rubber bands

pencils & pens

Robo-Cup

Robo-Dog

Shaker-Car

SNEAKY VIBRA-BOAT

Just as the vibrating motor can move the mini robot along a solid
surface, you can propel a small watercraft in the same fashion. Instead
of a cup, a sponge is used for the body of the watercraft.

Surface tension is one of the properties of water that causes it to
form into droplets. Water molecules have a sticky nature and they tend
to bind together. It's why, if you carefully pour water into a glass, it
will rise slightly over the top without spilling. It also allows you to float
carefully placed items (e.g., a paper clip) on its surface without sinking.

This project will make an intriguing display of the strength of water
surface tension, using a motorized water vehicle without a propeller.

Note: You should never mix electricity and water! But, don't worry,
the single AA battery used in this project only produces 1 1/2 volts of
electricity and is safe to use in water.

What's Needed

sponge pad

> Sponge or aluminum scouring pad
> Scissors
> Sink, bucket, or tub of water
> Mini-motor
> Large paper clip
> Three-volt watch battery
> Electrical tape

3-volt watch
battery

mini-motor

paper clip

scissors

electrical
tape

What to Do

Cut a 4-inch square piece of sponge, as shown in **Figure 1**. Ensure that the sponge is dry and test it to ensure it floats. See **Figure 2**.

Next, construct the motor and battery set, as shown in the previous project. This time, place the parts on top of the sponge inside the lid. Mount the motor high enough so its paper clip can move freely, as shown in **Figure 3**.

Last, carefully set the sneaky motorboat on the water until it floats. Gently tape the loose wire to the battery and let go. The little boat will sail on its own course in no time, as shown in **Figure 4**.

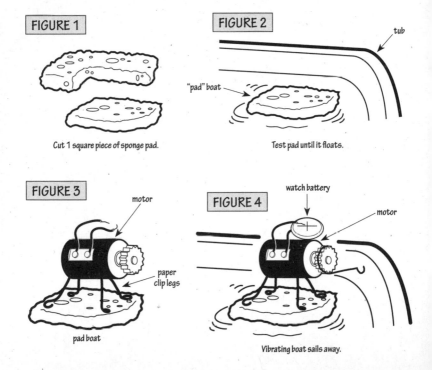

FIGURE 1

Cut 1 square piece of sponge pad.

FIGURE 2

tub

"pad" boat

Test pad until it floats.

FIGURE 3

motor

paper clip legs

pad boat

FIGURE 4

watch battery

motor

Vibrating boat sails away.

SNEAKY TRASHFORMER ROBOT

Product packaging incorporates special shapes to use inexpensive, thin cardboard to hold larger volumes of weight and to be able to stack them without being crushed. To demonstrate this, you can start with ten flat pieces of cardboard and produce a very sturdy "trashformer" robot.

The robot has an easy-to-build, adaptable design constructed from discarded product packaging (with a radio-controlled truck or car chassis as its base). This sneaky robot is not only made from recycled items, it also promotes recycling, since it includes pockets to collect discarded soda cans, small bottles, and batteries.

For adaptability, sections are connected with Velcro dots and rubber bands, and extra sections can be added when desired. Its head and arms are extendable and include blinking LED flashers for nighttime fun. The torso can be folded over, using a discarded CD or DVD case, and covered with cloth to act as a motorized meal tray. Additional accessories, such as a walkie-talkie or cell phone, can be attached with Velcro for added utility. The robot can be decorated and painted. Choose a medium-size vehicle with thick tires and a not-too-fast motor, for stability. (Some high-torque toy vehicles accelerate too fast and they can cause the robot to tip over.)

If you plan to have 'bot battles with friends, choose remote-control vehicles on different radio frequencies so they don't interfere with one another. You'll find the information on the outside of the box, usually 49 MHz or 27 MHz or channel numbers.

What's Needed

- Chassis from a radio-controlled truck or car
- Screwdriver
- Flexible antenna wire
- Velcro dots
- Plastic CD or DVD "jewel" case
- Four large cereal boxes /small
- Transparent tape
- Large rubber bands (optional)
- Two long cookie boxes
- Two toothpaste boxes (slightly smaller than the cookie boxes) make rubber bands
- Popcorn box make
- Three tea boxes make
- Long rectangular bag
- LED blinking lights, LED clip-on, or flexible reading lights (optional)
- Scissors

2 long cookie boxes

car chassis

cd case

velcro dots

screwdriver

flexible antenna wire

tape

led light

popcorn box

2 toothpaste boxes

long rectangular bag

scissors

3 tea boxes

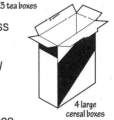
4 large cereal boxes

What to Do

When paper is folded, it resists bending due to stress distribution. When it's a flat sheet, paper has very little bending resistance. It can be easily bent in any direction. Once a sheet of paper is bent or curled, however, it is able to resist bending. At right angles to the bend it shows maximum strength. A rolled piece of paper becomes a strong pillar, or column.

The sneaky robot you'll make will be constructed with product packaging boxes that have also been designed to maximize their load-carrying capacity.

First, test the toy vehicle's ability to accelerate forward and backward, and its turning capabilities, before you dismantle it and remove the chassis. Even new or little-used cars may have problems and retail stores may not honor the warranty afterward.

Turn over the car and remove the screws that hold the truck body to the chassis. Carefully lift the body off, pulling the flexible antenna wire through the chassis hole, if necessary. Leave the front bumper/winch assembly on the front to provide extra stability for front-end bumping. Save the body parts and screws.

Apply four Velcro dots to the top of the car chassis and to the bottom of two cereal boxes. See **Figure 1**. Mount the boxes on the chassis with the front slightly apart to resemble legs. (Stuff extra folded cardboard between the lower front of the boxes so they are divided and resemble legs standing apart and a slight angle.) Apply four Velcro dots to the top of the cereal boxes and four more dots to the top and

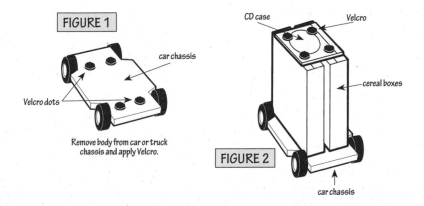

FIGURE 1

car chassis

Velcro dots

Remove body from car or truck
chassis and apply Velcro.

CD case   Velcro

cereal boxes

FIGURE 2

car chassis

bottom of the CD case. Place the CD case on the top of the two cereal box "legs." Secure the lower part of the CD case to the boxes and to the chassis. This will stabilize it when it accelerates and turns. See **Figure 2**. Apply Velcro dots to the top, bottom, and sides of the remaining two cereal boxes. Mount the cereal boxes on top of the CD case. See **Figure 3A**.

You can secure the cereal boxes in various ways, depending on whether you always want to keep the robot in its full-size form, or want to tilt it over half-height to double as a radio-controlled tray. Or, you may prefer to completely flatten all of the boxes so it can be easily transported in a large backpack or stored away on a shelf. Depending on your design, you can stabilize the upper and lower cereal boxes with tape or, for a foldable design wrap, use large rubber bands around the top two boxes and another pair of rubber bands to secure the lower boxes to the car chassis.

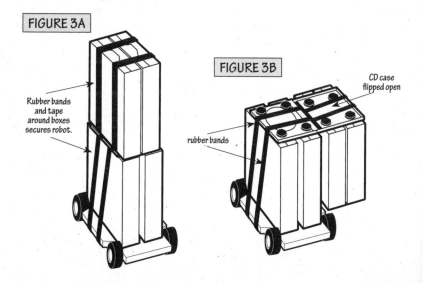

FIGURE 3A

Rubber bands
and tape
around boxes
secures robot.

FIGURE 3B

CD case
flipped open

rubber bands

For instance, tape the two boxes together at the top and attach the rubber band to it and the top door of the CD case. The top section should be able to freely swing over and downward so, when covered with cloth, it can act as a motorized meal tray. See **Figure 3B**.

To make the robot's arms, press Velcro dots to the sides of two long cookie boxes and slide one toothpaste box into each cookie box. Leave the flaps open and bend them curved on the toothpaste boxes so they resemble claws, as shown in **Figure 4**. The two boxes should stay together with friction. If there is a big difference in their sizes, you can stuff small pieces of paper between them to ensure a tight fit.

To make the robot head, using the popcorn box (Box 1), cut a triangular or trapezoid-shaped hole near the top front of the box to resemble a robotic Cylops eye. Then slide the head section over the tea box (Box 2) that will act as a neck assembly with Velcro dots on the bottom. See **Figure 5**.

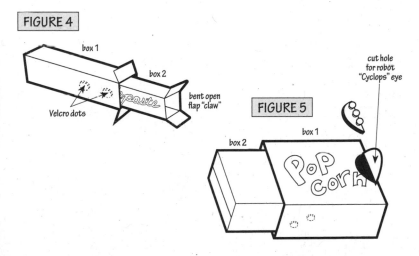

FIGURE 4

box 1

box 2

cut hole for robot "Cyclops" eye

bent open flap "claw"

Velcro dots

FIGURE 5

box 1

box 2

PoP Corn

You can draw a battery and a recycling symbol (a triple arrow) on the front of the bag (which you should tape to the front of the robot), as shown in **Figure 6**.

Last, tape or press on the robot's arms and neck and extend them as desired, as shown in **Figure 7**.

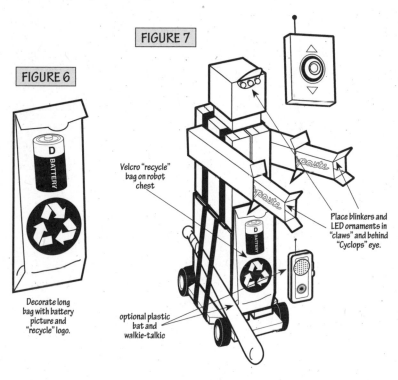

FIGURE 7

FIGURE 6

Velcro "recycle" bag on robot chest

Place blinkers and LED ornaments in "claws" and behind "Cyclops" eye.

Decorate long bag with battery picture and "recycle" logo.

optional plastic bat and walkie-talkie

ACCESSORIZING THE ROBOT

The boxy shapes of the robot's body parts can be enhanced with a few add-ons. You only need a few household items to produce a more dramatic "mechanized" appearance.

What's Needed

▶ Cardboard product packaging from 3 cereal boxes and 2 rectangular tea boxes

▶ Red light flasher, sequential type if available

▶ Batteries for the flasher

▶ Tape

▶ Velcro

▶ Scissors

▶ Flexible mini LED lights

▶ Flexible plastic tubing

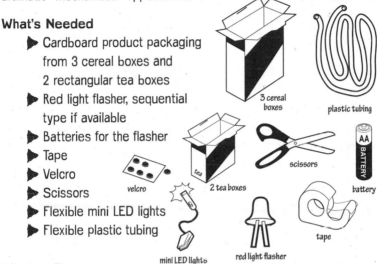

3 cereal boxes

plastic tubing

tea

2 tea boxes

velcro

scissors

mini LED lights

red light flasher

tape

battery

What to Do

All of the decorations described here can be modified for your particular robot's size and shape. The add-on helmet, leg shields, and so on can be reshaped as well as adorned with stickers and other add-ons. And, with careful planning and placement, you can fold up the torso and leg cereal boxes completely flat, with the add-on boxes also folded in place. This allows you to quickly disassemble and store the entire robot in a small backpack. You can also transport the robot and reassemble it in a matter of minutes by unfolding the pieces and adding a few pieces of tape.

HEAD GEAR

For a more imposing look, cut out the winged helmet pattern from a piece of cardboard. Tape it to the front and side of the head section. See **Figure 1**.

With a light flasher, especially the sequential type, you can add a futuristic "Cylon" look to your homemade robot. Test the flasher with fresh batteries, and tape or Velcro it in place behind the eye cutout near the top of the head, as shown in **Figure 2**.

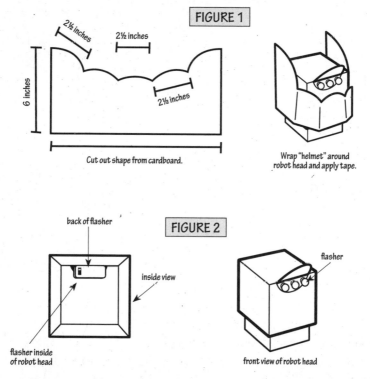

FIGURE 1

2½ inches

2½ inches

6 inches

2½ inches

Cut out shape from cardboard.

Wrap "helmet" around robot head and apply tape.

back of flasher

FIGURE 2

inside view

flasher inside of robot head

flasher

front view of robot head

SHOULDER GUARDS

By attaching two tea boxes, with tape or Velcro, you add girth to the robot's appearance. **Figure 3** shows a tea box mounted on top of the back of the cookie box arm section. Apply Velcro to the side of the tea box and to the robot's upper cereal box torso section to secure the shoulder and arm sections.

Clip the flexible light to the top of the tea boxes (for the left and right arm and shoulders) and slide the tubing between the front of the cookie and toothpaste boxes and into the top of the tea boxes, as shown in **Figure 4**.

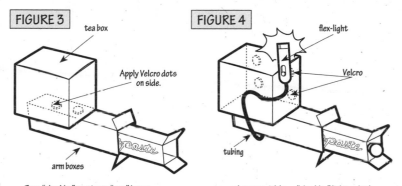

FIGURE 3

tea box

Apply Velcro dots on side.

arm boxes

Tape "shoulder" section to "arm" boxes.

FIGURE 4

flex-light

Velcro

tubing

robot arm with large "shoulder" light and tube

LEG SHIELDS

Add protection for robot's lower legs with the easy-to-make shields cut from spare cardboard. From a cereal box, cut a trapezoidal cardboard 7 inches long, 6 inches wide on the top and 8 inches wide on the bottom. Fold it into three equal sections (at the top there will be three 2-inch folded sections). See **Figure 5**. Tape the sides of the cardboard to the sides of the lower robot boxes to add an imposing leg shield, as shown in **Figure 6**.

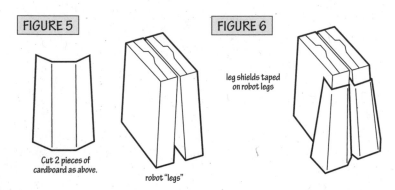

FIGURE 5

Cut 2 pieces of cardboard as above.

robot "legs"

FIGURE 6

leg shields taped on robot legs

With the suggested additions to the robot, including a recycling bag taped to the front (to collect used batteries) and more flexible lights clamped to the back of the robot's torso, it will exhibit an imposing look. Compare the original design to the fully adorned robot in **Figure 7**.

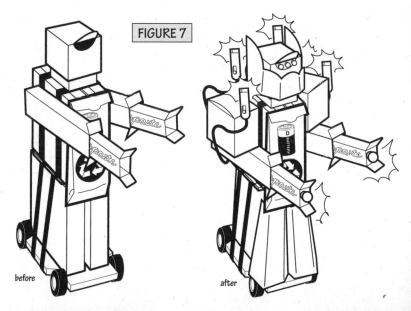

FIGURE 7

before

after

SNEAKY MEASUREMENT PROJECTS

Think about your typical day: you might not realize how many times you measure something. Measurement is the process of describing the characteristics of things, using numbers. For example, when you cook, you measure out ingredients, mix items together, and adjust the cooking temperature and time settings. When you step on a scale, you are measuring your weight. When you check how long a board is, with a measuring tape, you are measuring length. In this section you'll learn how to make sneaky instruments such as an altimeter, a barometer, and a voltmeter that can measure things that you cannot reach, feel, or see but that are nonetheless important.

Did you know you can use water to find out your elevation? Or use string, a drinking straw, and cardboard to tell the height of a faraway building? Turn a piece of wire into an electrical voltmeter? Or use plastic wrap to detect the air pressure in a room? All of these feats are possible if you think sneakily.

SNEAKY WEATHER BAROMETER

Before modern weather satellites and radar systems could precisely measure the weather, people depended on simple barometers to measure air (or barometric) pressure. Using everyday items, you can make a barometer to give you a heads-up about weather conditions.

What's Needed

- Plastic wrap
- Scissors
- Wide-mouthed jar or large plastic cup
- Large rubber band
- Lightweight drinking straw
- Tape
- White cardboard
- Pen

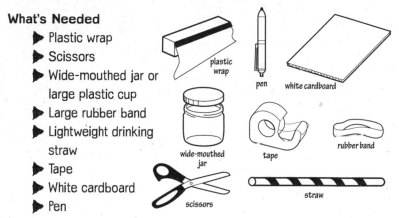

What to Do

Cut a piece of plastic food wrap large enough to cover the top of the jar and fold it over, halfway down the sides. Pull on the sides of the plastic wrap for a drum-tight fit—to work, this project must have an airtight seal. Wrap the rubber band around the side of the jar to secure a tight fit. See **Figure 1**.

FIGURE 1

Place the straw on top of the plastic wrap so it rests horizontally on the surface with its end near the middle of the jar opening, as shown in **Figure 2**. Use a small piece of tape to secure in place.

Stand the cardboard behind the jar and note the height of the straw. Draw horizontal lines in fine increments on the cardboard, above and below the height of the straw, as shown in **Figure 3**.

Place the sneaky barometer in a location where there are no drastic temperature changes (e.g., not near a radiator or window). Note the position of the straw to see changes in the straw's position, especially before and after a changing local weather condition. Graph the changes on the cardboard gauge behind the straw, for future reference. See **Figure 4**.

FIGURE 2

Tape straw to top of jar.

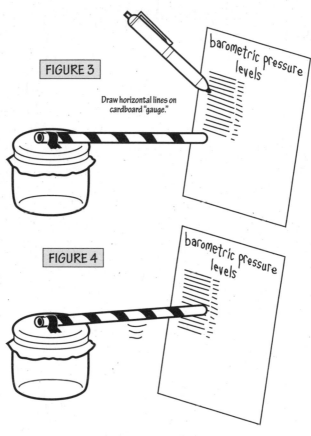

FIGURE 3

Draw horizontal lines on
cardboard "gauge."

barometric pressure
levels

FIGURE 4

barometric pressure
levels

HOW IT WORKS

Air pressure is pushing in every direction all around us. During dry
and calm weather—a high atmospheric pressure condition—it presses
against the top of the jar, making the other end of the straw rise.

Conversely, approaching wind and rainstorms are usually preceded by
a low-pressure condition. The lowering straw level is an indicator of this.

SNEAKY ALTIMETER

Air pressure is actually a measurement of the weight of all the atmosphere above us, so it follows that there is more air pressure at low altitudes (with more air above us) than at higher ones. A device that measures air pressure at different altitudes is called an *altimeter*. Just as a barometer measures changes in air pressure, using a sealed jar and straw, an altimeter allows you to gauge your elevation level by using a tube of water.

What's Needed

- 3 by 1-foot piece of cardboard
- Twelve feet of clear, flexible tubing
- Funnel or turkey baster
- Tape
- Water
- Paper
- Pen
- Scissors

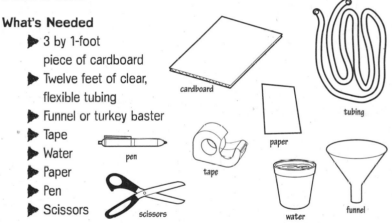

cardboard

tubing

pen

tape

paper

water

funnel

scissors

What to Do

Fold back the sides of the cardboard sheet by 2 inches to form supports on the sides so it can stand erect, as shown in **Figure 1**.

Mount the tubing on the cardboard with tape. Be sure to leave a longer length of tubing on the left side, coiled up and bent closed at the top. Wrap tape around this end to seal it shut. See **Figure 2**.

Next, use the funnel to pour water in the right side of the tube, until the water levels on both sides rise to half the length of the cardboard. See **Figure 3**. The funnel is needed to prevent the buildup of water bubbles inside the tube that would occur if you tried to pour in the water from an open faucet.

Draw multiple horizontal lines on the paper and tape it behind the tubing, as shown in **Figure 4**. This will act as a graphical indicator of altitude. The water level on both sides of the tubing is the same because there's equal pressure of air above water.

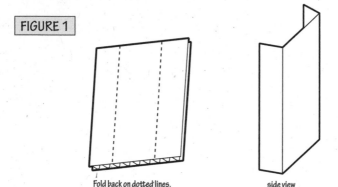

FIGURE 1

Fold back on dotted lines.

side view

Air pressure on both ends of the tube is at ground level (or very near sea level). Your altimeter measurements will be relative to the air pressure where you filled the tube with water and sealed off the left tube end. Any changes in altitude, whether the altimeter ascends or descends, will be indicated accordingly by a difference in the water level on the left or right side of the tube. Mark the height of the water on the scale you've drawn, then carefully take the sneaky altimeter to a

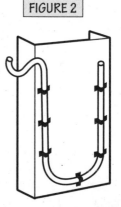

FIGURE 2

higher altitude in your area (e.g., at the top of a hill or the top floor of a tall building). Avoid letting the water spill out of the tubing.

You will notice that the higher you ascend, the lower will be the water level on the left side, and the right side will rise. This is because there is less pressure on the open side of the tube to push down on the water, compared to the air on the right side, which was trapped inside while the altimeter was constructed at ground level. See **Figure 5**.

As you change altitude, you can mark the lines on the paper in relation to the water level, to indicate the height, or present altitude. Also, if you know the height of the hill or mountain you're on, mark that height on the scale. You now have made a sneaky altimeter!

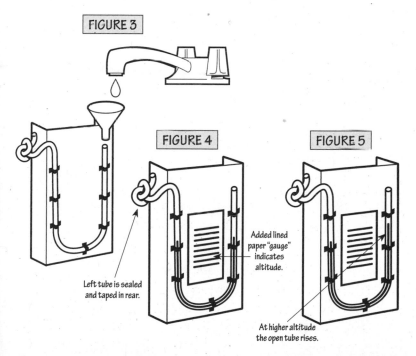

FIGURE 3

FIGURE 4

FIGURE 5

Left tube is sealed and taped in rear.

Added lined paper "gauge" indicates altitude.

At higher altitude the open tube rises.

MAKE AN ANEMOMETER

In addition to barometric pressure, another useful piece of weather measurement is wind speed, which is measured using an anemometer. The first mechanical anemometer was invented in 1450 by Italian artist and architect Leon Battista Alberti. Anemometers are now used in windmill guidance systems, to rotate them in the best direction to take advantage of the prevailing wind direction.

You can make a simple anemometer with everyday items and calibrate it on your next passenger ride in a car (with someone else driving, of course!).

What's Needed

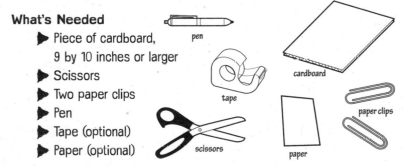

- Piece of cardboard, 9 by 10 inches or larger
- Scissors
- Two paper clips
- Pen
- Tape (optional)
- Paper (optional)

pen

tape

scissors

cardboard

paper clips

paper

What to Do

Cut a 1/2 by 9-inch strip from the cardboard sheet. Then, cut off 2 inches from the strip so it measures 1/2 by 7 inches long. This leaves the remaining piece of cardboard in the shape of a rectangle. See **Figure 1**.

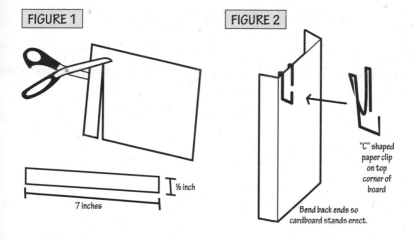

FIGURE 1

½ inch

7 inches

FIGURE 2

"C" shaped
paper clip
on top
corner of
board

Bend back ends so
cardboard stands erect.

Bend back the sides of the cardboard by 1 inch to allow it to stand up vertically. Next, bend the two paper clips into a **C** shape and push one of them through the top left corner of the cardboard, as shown in **Figure 2**. This will act as a support mount for the anemometer's indicator strip. Wrap the paper clip from the back of the cardboard over the top, and press it tightly so it doesn't move.

Fold over one end of the cutout cardboard strip 1 inch from the end and place it over the paper clip attached to the cardboard, as shown in **Figure 3**.

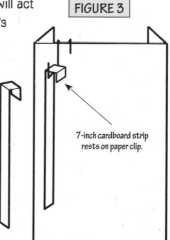

FIGURE 3

7-inch cardboard strip
rests on paper clip.

Push the other paper clip through the cardboard sheet at the lower left corner of the cardboard, about 3 inches from the bottom, to act as a rest stop. See **Figure 4**.

Now when the wind blows, turn the sneaky anemometer in the direction of the wind and the cardboard strip will move. Draw angled lines on the cardboard, or on a piece of paper taped on the surface, which provides flexibility in its position and design, which resembles the lines of a protractor. See **Figure 5**.

Last, to calibrate your anemometer, have a friend drive you in an open area and carefully hold the anemometer outside the window at various speeds. Mark the paper gauge on the cardboard to indicate the speeds.

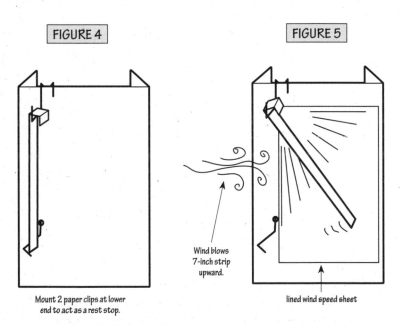

FIGURE 4

FIGURE 5

Wind blows
7-inch strip
upward.

Mount 2 paper clips at lower
end to act as a rest stop.

lined wind speed sheet

MAKE A HYPSOMETER

Look out the window at a tall building or structure that is some distance away. How tall is the building? You could guess, but there is a better, sneakier way to find out.

You can estimate the height of objects that are too far away or too tall to physically measure, with an easy-to-make device called a hypsometer (*hyps* means "height" in Greek). When you know the distance from an object and the angle at which to see its highest point, you can use a simple principle of trigonometry to calculate its height.

Armed with this knowledge, you can make a sneaky hypsometer with a drinking straw, cardboard, and a paper clip.

What's Needed

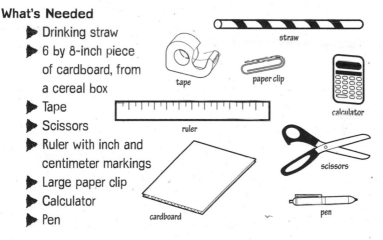

- Drinking straw
- 6 by 8-inch piece of cardboard, from a cereal box
- Tape
- Scissors
- Ruler with inch and centimeter markings
- Large paper clip
- Calculator
- Pen

straw

tape

paper clip

calculator

ruler

scissors

pen

cardboard

What to Do

Place the straw 1 inch from the top of the cardboard and tape it securely. Then, puncture a small hole in the right side, 10 centimeters from the bottom. Straighten the paper clip and bend one end. See **Figure 1**.

Roll the top section of the cardboard over the straw and tape it in place, as shown in **Figure 2**. Hook the paper clip into the hole. The paper clip should be able to hang freely and swing back and forth. See **Figure 3**.

Next, write numbers from zero to twelve, 1 centimeter apart from right to left, across the bottom of the board, with the zero starting at the hanging paper clip and proceeding leftward to number 12. See **Figure 4**.

To test the hypsometer, note the distance you are from another person. For this example, hold the cardboard with one hand at an angle and look through the straw to view the top of a friend who is positioned exactly 10 feet away. You'll notice that when you tilt your head (and the

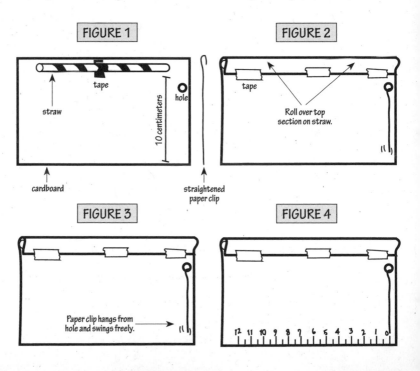

position of the board), the paper clip will move leftward at an angle and point down at the floor or ground. Check to see which number the paper clip points to. In this example, it is number 1.

With the distance in the number of centimeters you are from the object, and the angle of tilt shown on the hypsometer, you're ready to calculate your friend's height.

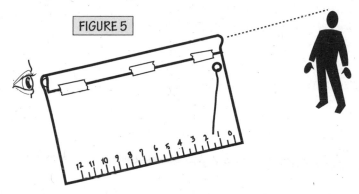

FIGURE 5

Note: Metric measurements are much easier to calculate with a hypsometer because it uses a base 10 decimal system. The inches/feet/yards system isn't good for scientific calculations. Many Web sites offer standard-to-metric calculators. If you don't have a metric ruler handy, you can use a calculator and the following chart to help you with your estimates:

FIGURE 6

1 inch = 2.5 centimeters
1 foot = 30 centimeters
3.3 feet = 1 meter
1,000 meters = 1 kilometer (sometimes called "clicks" in the military)
1 millimeter = 0.001 meter = 0.04 inch
1 centimeter = 0.01 meter = 0.4 inch

You should write these figures on the back of the hypsometer, for easy reference.

You'll need two more figures for your sneaky height estimate: the height of your eye line, from the floor to your eyes, and the height of the card. Fortunately, these two measurements don't change, making it easy to add them to the total. Simply measure your eye line from the top of your head, which is usually 3 to 5 inches, and subtract this number from your height. For instance, if you are 5 feet, 8 inches tall and your eye line is 4 inches from the top of your head, the eye height is 5 feet, 4 inches, or 164 centimeters. The height of the card is 10 centimeters. With these figures, you're ready to estimate your friend's height with the hypsometer.

In this example, multiply the distance to your friend in centimeters (305) by the number that the paper clip pointed to (1), then divide by the height of the card in centimeters (10). Add this number to your eye height in centimeters (164), for a total of 194.5 centimeters.

$$305 \text{ cm} \times 1 = 305 \text{ cm}$$
$$\div 10$$
$$= 30.5 \text{ cm}$$
$$+ 164.0 \text{ cm}$$
$$= 194.5 \text{ cm}$$

You can convert this number to standard feet and inches, approximately 6 feet, 3 inches tall.

Test your hypsometer with objects of unknown heights, such as doors, walls, and buildings, to calibrate the hypsometer and sharpen your estimation skills.

SNEAKY VOLTMETER

Voltmeters use the principle of magnetic attraction to indicate the level of electricity. A coil of wire with electricity flowing through it will become an electromagnet and attract a piece of metal nearby.

If the piece of metal is attached to a movable object, it will act as an electrical power indicator. If a piece of magnetized metal is balanced near a coil of wire, it can be used to indicate the relative voltage flowing through the wire.

What's Needed

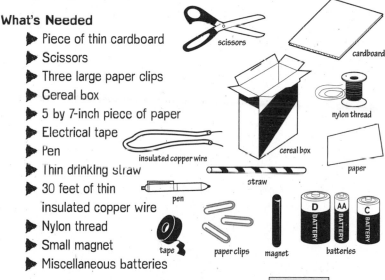

- Piece of thin cardboard
- Scissors
- Three large paper clips
- Cereal box
- 5 by 7-inch piece of paper
- Electrical tape
- Pen
- Thin drinking straw
- 30 feet of thin insulated copper wire
- Nylon thread
- Small magnet
- Miscellaneous batteries

What to Do

First, cut a piece of thin cardboard to be 7 inch by 1/2 inch, with one end cut on a slant. See **Figure 1**. This will act as the pointer for the voltmeter.

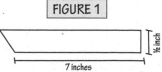

FIGURE 1

Cut a 7-inch by ½-inch length of cardboard and cut a pointed shape on the left side.

Next, bend the end of a paper clip so it points outward and tape it to the top center of the cereal box, as shown in **Figure 2**.

Tape the piece of paper to the upper left corner of the cereal box. Draw indicator lines that start in the left center and curve to the top. See **Figure 3**.

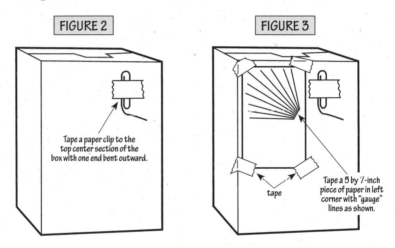

FIGURE 2

Tape a paper clip to the top center section of the box with one end bent outward.

FIGURE 3

tape

Tape a 5 by 7-inch piece of paper in left corner with "gauge" lines as shown.

Next, puncture a small hole in the center of the cardboard pointer and slide it onto the paper clip, as shown in **Figure 4**.

Cut off a 2-inch length of drinking straw and wrap two hundred turns of wire around it, in one direction. To do this, first leave a 5-inch length of wire at the bottom end, then start winding the coil tightly toward the other end. See **Figure 5**. If you reach the top of the straw before you get to two hundred turns, pull the wire down to the bottom and start winding upward toward the top until it's complete. Apply tape to keep the wire tight and secure as needed. Also leave another 5-inch length of wire free at the bottom. Apply more tape to secure the coil, and strip the insulation from the ends of the wire. See **Figure 6**.

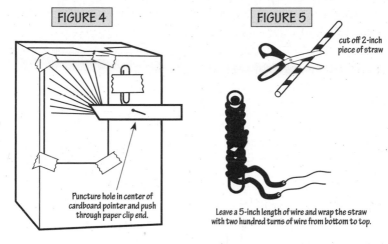

FIGURE 4

FIGURE 5

cut off 2-inch piece of straw

Puncture hole in center of cardboard pointer and push through paper clip end.

Leave a 5-inch length of wire and wrap the straw with two hundred turns of wire from bottom to top.

Next, make a small hole in the lower right end of the pointer. Pull the thread through the hole and tie a secure knot. Then straighten one of the paper clips and bend a small loop at one end. Tie the paper clip loop through the thread so it hangs down about 4 inches from the pointer. Slide a second paper clip on the upper left edge of the pointer so it balances the weight and the pointer sits level. See **Figure 7**.

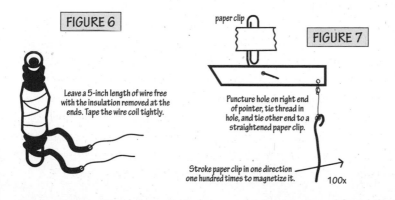

FIGURE 6

paper clip

FIGURE 7

Leave a 5-inch length of wire free with the insulation removed at the ends. Tape the wire coil tightly.

Puncture hole on right end of pointer, tie thread in hole, and tie other end to a straightened paper clip.

Stroke paper clip in one direction one hundred times to magnetize it.

100x

Next, straighten a third paper clip and leave one end with a curved loop. Rub the magnet across the surface of it in one direction only, one hundred times, so it becomes magnetic. Then, slip the end of the paper clip into the center of the coil so that the end is still visible. Tape the coil securely to the front of the cereal box. The tester is ready. The coil and magnetic paper clip become a solenoid, a device that causes motion when a current of electricity flows through a wire. In this example, current flowing through the wire coil attracts the magnetized paper clip and pulls it downward. This, in turn, tugs the right side of the cardboard pointer and its left side moves upward, indicating a measurement on the paper meter.

Press the wire ends to the positive and negative terminals of a battery. This will send electricity flowing through the coil, which attracts the paper clip and pulls it downward. The left end of the pointer moves upward, indicating a voltage level on the paper gauge, as shown in **Figure 8**.

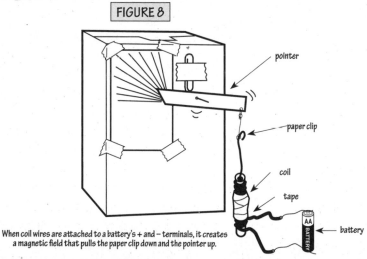

FIGURE 8

pointer

paper clip

coil

tape

battery

When coil wires are attached to a battery's + and − terminals, it creates a magnetic field that pulls the paper clip down and the pointer up.

SNEAKY ASTRONOMY AND NAVIGATION TRICKS

The Sun, Moon, and (sometimes) stars can be seen anywhere on Earth and they can be used to give you clues about where you are and where you're going.

In ancient times, the practical need for timekeeping and navigation were two of the primary reasons for the study of astronomy. The celestial origin of our timekeeping and navigation systems is still evident. The time of day comes from the location of the Sun in the local sky, months are derived from the Moon's cycle of phases, and the year comes from the Sun's annual circle among the stars. The devices used for timekeeping, however, have evolved from the sundial to advanced atomic clocks.

Celestial navigation is the art and science of finding your way by the Sun, Moon, and stars. People have been determining direction, finding routes and locations, and orienting themselves from the positions of celestial bodies since the beginning of history because the stars are (relatively) fixed markers. With the rise of radio and electronic means of finding location, knowledge of celestial navigation has experienced a decline.

The techniques and projects that follow allow you to make easy-to-assemble instruments, using a paper clip, cardboard, a wristwatch, a pencil, and sometimes just a sneaky sense of observation of the sky to find your directions, night or day.

MAKE A SNEAKY COMPASS

To locate magnetic north—not the North Pole but close—you need a compass. You can quickly make a sneaky one with just a paper clip and magnet (or you can replace the magnet with clothing if it is made of silk or a synthetic fiber).

What's Needed
▶ Paper clip
▶ Magnet
▶ Pen cap

magnet   paper clip

pen cap

What to Do
You can obtain magnets from a variety of everyday things including an old radio, headphones, tape recorder speaker, shaker-style flashlight, or some hearing-aid battery packages. Small motors inside toy cars and other devices also have magnets inside. Rubbing the motor case against a paper clip multiple times will magnetize it. Micro radio-control cars include a tiny magnet between the front wheels, for steering, that can be salvaged for sneaky projects. See **Figure 1**.

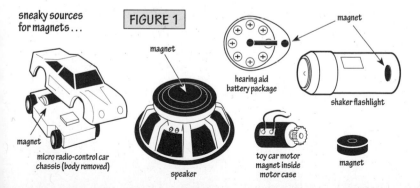

sneaky sources for magnets . . .

FIGURE 1

magnet

hearing aid battery package

magnet

shaker flashlight

magnet

micro radio-control car chassis (body removed)

speaker

toy car motor magnet inside motor case

magnet

First, straighten the paper clip and bend it into the shape shown in **Figure 2**. Next, stroke the long end of the paper clip with the magnet in one direction, not back and forth. This will align the paper clip's electrons in a uniform direction. See **Figure 3**.

Last, balance the end of the paper clip on top of a pen cap, as shown in **Figure 4**. You'll see that it will slowly swing to the north or south direction. It if actually points south, you must stroke the paper clip in the other direction next time.

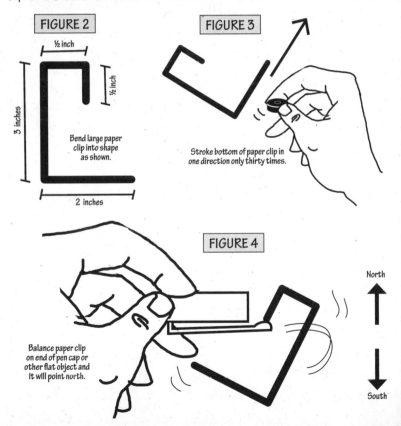

FIGURE 2

½ inch

½ inch

3 inches

Bend large paper clip into shape as shown.

2 inches

FIGURE 3

Stroke bottom of paper clip in one direction only thirty times.

FIGURE 4

Balance paper clip on end of pen cap or other flat object and it will point north.

North

South

SNEAKY SUNDIALS

Sundials measure time, based on the actual position of the Sun in the sky. This time is called the *apparent* (or local) *solar time*. Noon is the precise moment when the Sun is on the meridian (which is an imaginary line passing from north to south through the zenith, the point directly overhead) and the sundial casts its shortest shadow. Before noon, when the Sun is on its way to the meridian, the apparent solar time is antemeridian (or a.m.), meaning before the meridian; and past noon, the apparent solar time is postmeridian (p.m.), thus the abbreviated terms we use.

Before people used windup, analog, or digital clocks, the Sun was the main means of telling time. In the past it wasn't unusual for people to carry a foldable sundial, containing a small compass to locate the north direction.

With a few ordinary items you can make a sneaky timepiece that you can use at home and while traveling.

Traditional Sundial

What's Needed
▶ Protractor
▶ Pen
▶ Cardboard
▶ Scissors
▶ Pencil or long wooden stick

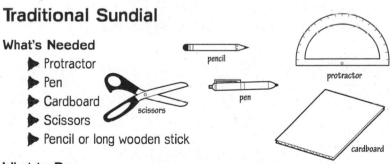

pencil

protractor

scissors

pen

cardboard

What to Do
On a sunny day, any vertically standing stick can act as a crude sundial. There is one problem to overcome, though—the sun shines at different angles at different latitudes on Earth. To increase the accuracy of your

sneaky sundial, the pointer must be slanted at the same angle as the latitude of your location.

First, check the latitude list at the end of the book and find the city closest to you. (Or, check online for the latitude of your exact location.) Subtract your latitude from 90 and note this number. For example, if you live in or near Salt Lake City, which has a latitude of 40 degrees, when you subtract this number from 90, the difference is 50 degrees.

Then, use the protractor to draw two right-angle triangles on the cardboard with a 50-degree opposite angle, as shown in **Figure 1**. Position another square piece of cardboard so it creates an angled stand for your sundial. See **Figure 2**.

FIGURE 1

50°   40°

50°

90°   cardboard with 50° slant

Next, draw on the front of your sundial stand a semicircular scale that resembles the one on your protractor. Cut a small hole through the center of the stand and push the pencil through it. See **Figure 3**. On a sunny day, use a compass (your sneaky one will serve perfectly) to position the sundial so that the cardboard is parallel to the east/west direction.

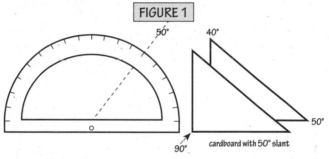

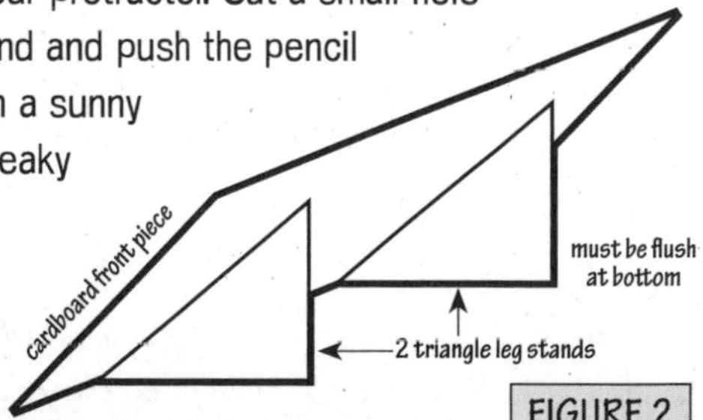

cardboard front piece

must be flush at bottom

2 triangle leg stands

FIGURE 2

Note and mark the position of the shadows that appear at each hour, and you have a free solar timepiece. Once you're familiar with the proper latitude angle, and the east/west position of your present location, you can make an improvised sundial out of all sorts of everyday objects, including a soda cup and your fingers.

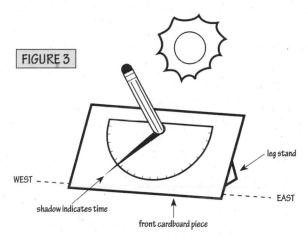

FIGURE 3

WEST - - - -

EAST

shadow indicates time

front cardboard piece

leg stand

Sneaky Cup and Straw Sundial

What You Need

▶ Large, wide-mouthed cup with lid

▶ Long drinking straw

▶ Pen

straw

pen

large cup with lid

What to Do

You can create a makeshift sundial with a large, wide-mouthed cup and straw. First, place the straw through the hole in the cup lid. Next, line up the cup in the east/west direction and position the straw at the proper angle until you see a shadow appear on the surface of the cup lid. The shadow that appears will move from left to right. Mark the lid with a pen at the top of each hour to create your sundial. See **Figure 1**.

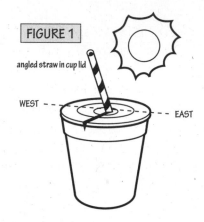

FIGURE 1

angled straw in cup lid

WEST – – – – – – – – EAST

Sneaky Hand and Pencil Sundial

What You Need

▶ Pencil, stick, or pen

pencil

What to Do

With enough practice, you can position a stick or writing instrument in your hand and attain a fairly accurate reading of the time. On a sunny day, turn your body westward before noon or eastward after noon. As stated in the traditional sundial project, you must determine your latitude by referring to a globe, map, or online. Then, position the pencil in the position shown in **Figure 1**. The shadow on your hand determines the approximate hour of the day. This handy method also works with a pen or a straight stick.

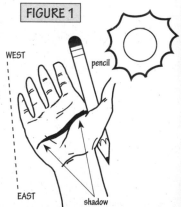

FIGURE 1

WEST

pencil

EAST

shadow

Bonus Trick: Sneaky Moon Direction Trick

In many urban areas you cannot see the stars above because of smog and the scatter of city lights, but usually the Moon is still visible. Here's a sneaky method to tell your direction with the Moon.

Look at the Moon and if it is in a crescent phase, with two points visible, you're in luck. Simply imagine a line extending from the two points leading down to earth. That direction is south (or nearly south). See **Figure 1**.

Naturally, the opposite direction is north, and you can locate the east and west directions with ease.

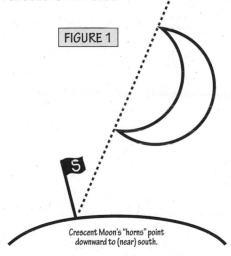

FIGURE 1

Crescent Moon's "horns" point
downward to (near) south.

SNEAKY QUADRANT

The sneaky hypsometer you created in the previous section isn't
just for height determinations! With it, you can make a quadrant, also
referred to as a *sextant or astrolabe*—a device that can help you detect
your present latitude, your position north or south of the equator, by
viewing the North Star.

How do you locate your position by observing a star? Just as you
used the hypsometer to look at an object through a tube and measure
the reading on the gauge, you can locate and look at the North Star
(Polaris, or the Pole Star) and read a different gauge to determine the
angle of height (declination angle). By measuring the angle of elevation
of the North Star, you can determine your present latitude.

What's Needed
▶ Hypsometer
(see page 71)
▶ Protractor
▶ Tape

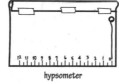

tape

hypsometer

protractor

HOW IT WORKS
Latitude (shown as a horizontal line on a map) is the angular distance,
in degrees, minutes, and seconds, of a point north or south of the
equator. Lines of latitude are often referred to as *parallels*. The equator
is 0 degrees latitude. The North Pole is 90 degrees.

Lines of longitude are often referred to as *meridians*. *Longitude*
(shown as a vertical line) is the angular distance, in degrees, minutes,
and seconds, of a point east or west of the prime (Greenwich) meridian
located in England.

Distance between Lines

If you divide the circumference of the Earth (approximately 25,000 miles) by 360 degrees, the distance on the Earth's surface for each single degree of latitude or longitude is just over 69 miles. At the North Pole, which is 90 degrees latitude, the North Star is directly overhead. At the equator, which is 0 degrees latitude, the star is on the horizon with 0 degrees altitude. Between the equator and the North Pole, its angle above the horizon is a direct measure of terrestrial latitude.

In ancient times, a navigator planning to sail out of sight of land would simply measure the altitude of the North Star as he left home port, in today's terms measuring the latitude of home port. To return after a long voyage, he needed only to sail north or south, as appropriate, to bring the star to the altitude of the home port, then turn left or right as appropriate and "sail across the latitude," keeping the North Star at a constant angle.

The mariner's quadrant—a quarter of a circle made of wood or brass—came into widespread use for navigation around 1450, though its use can be traced back at least to the 1200s. Using your sneaky quadrant, you can then take a measurement of the North Star and calculate your latitude.

Note: A protractor spans 180 degrees. It is easy to obtain and use for this project.

Which stars are important for navigation? There are several, but the most famous navigational stars are the North Star, and the Southern Cross. The North Star is a part of the constellation Ursa Minor, commonly known as the Little Dipper. The Southern Cross is a constellation of four stars called Crixa, two of which point toward the celestial South Pole.

What to Do

Puncture a hole near the top center of the hypsometer like the one on the right corner, as shown in **Figure 1**. Place the paper clip in it.

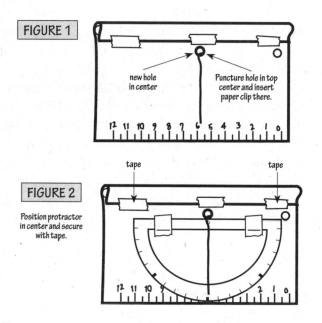

FIGURE 1

new hole in center

Puncture hole in top center and insert paper clip there.

12 11 10 9 8 7 6 5 4 3 2 1 0

tape   tape

FIGURE 2

Position protractor in center and secure with tape.

12 11 10   2 1 0

Then, position the protractor upside down and flipped over, so that the 0 is on the left-hand side. Tape the protractor to the center of the hypsometer so the paper clip points to 90 degrees. See **Figure 2**.

Your present latitude will be 90 degrees minus the angle indicated on the protractor by the paper clip. See **Figure 3**. For example, on a clear night, look at the stars until you can locate the Big Dipper constellation (also known as Ursa Major). The Little Dipper, Ursa Minor, is located just above and to the right of the Big Dipper. The Big Dipper's rightmost two stars point to a star on the "handle" of the Little Dipper.

That is the North Star. To be sure, look for the Cassiopeia constellation to the right of the Little Dipper. Cassiopeia forms a **W** shape and its lowest star also points to the North Star. (Together, the Big Dipper and Cassiopeia "bookend" the North Star.)

If you were located at the North Pole, the North Star would be directly overhead in the night sky. If you pointed your quadrant straight up to see it, the paper clip would swing to the 0-degree mark.

If you were at the equator and pointed the quadrant to look at the North Star, it would appear on the horizon and the reading on the protractor will be 90 degrees, since the paper clip would be hanging straight down. Halfway between the star and the equator, the protractor shows the paper clip at the 45 degree point. Los Angeles is 34 degrees (33.93) latitude, so when looking through the straw at the North Star (sometimes possible if you're outside the city area) the pointer will indicate approximately 56 degrees: 90 degrees minus 56 degrees = 34 degrees.

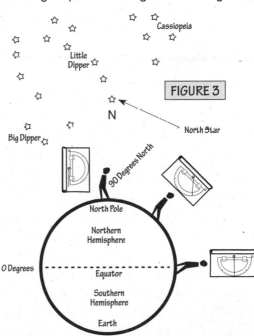

FIGURE 3

SNEAKY DIRECTION FINDING: USE A WATCH

If you're stranded without a magnetic compass, all is not lost. Even without a compass, there are numerous ways to find directions in desolate areas. Three methods are covered here.

What's Needed

> ▶ Standard analog watch
> ▶ Clear day where you can see the Sun

watch

What to Do
The Sun always rises in the east and sets in the west. You can use this fact to find north and south with a standard nondigital watch. If you are in the Northern Hemisphere (north of the equator), point the hour hand of the watch in the direction of the Sun. Midway between the hour hand and 12 o'clock will be south. See **Figure 1**.

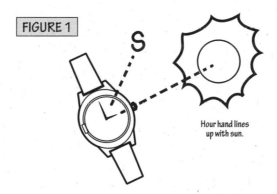

FIGURE 1

S

Hour hand lines
up with sun.

SNEAKY DIRECTION FINDING: USE THE STARS

What's Needed

▶ A clear evening when stars can be viewed

What to Do

In the Northern Hemisphere, locate the Big Dipper constellation in the sky; see **Figure 1**. Follow the direction of the two stars that make up the front of the dipper to the North Star. (It is about four times the distance between the two stars that make up the front of the dipper.) Then follow the path of the North Star down to the ground. This direction is north.

In the Southern Hemisphere, locate the Southern Cross constellation in the sky; see **Figure 2**. Also notice the two stars below the Cross. Imagine two lines extending at right angles, one from a point midway between the two stars and the other from the Cross, to see where they intersect. Follow this path down to the ground. This direction is due south.

FIGURE 1

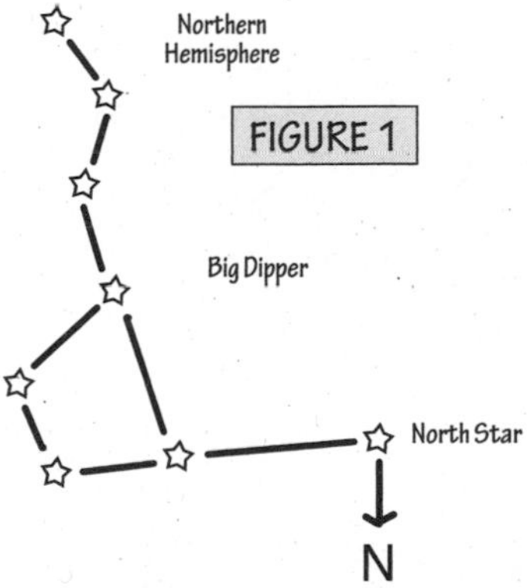

Northern Hemisphere

Big Dipper

North Star

N

FIGURE 2

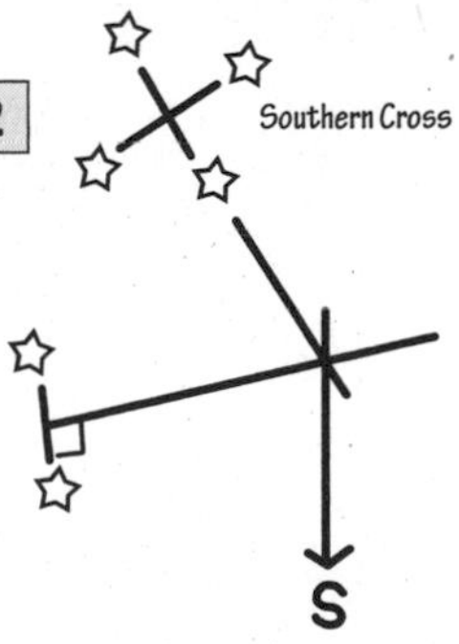

Southern Cross

S

SNEAKY DIRECTION FINDING: USE A STICK

What's Needed
▶ Stick or branch
about 3 feet long
▶ Rock or leaf

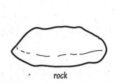

rock

sticks

What to Do
On a sunny day, you can find out which direction is north, south, east, or west by using shadows. Stand a stick upright in the ground, as shown in **Figure 1**. Notice the shadow it casts and, using a rock or leaf, mark the shadow's edge.

Wait about fifteen minutes and notice the new shadow that appears. Mark its tip, too. See **Figure 2**. Draw an imaginary line between the two marks. This is the east–west line (west is the first tip, and the second marker represents east). You can draw an imaginary or real line across the east–west line to determine the north and south directions. See **Figure 3**.

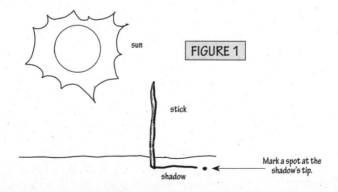

sun

FIGURE 1

stick

Mark a spot at the shadow's tip.

shadow

15 minutes later...

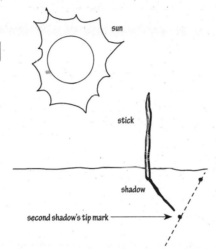

FIGURE 2

sun

stick

shadow

second shadow's tip mark

FIGURE 3

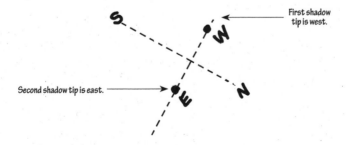

First shadow tip is west.

Second shadow tip is east.

SNEAKY MAGIC TRICKS

Magicians perform a wide variety of tricks with commercial gimmicks and stage devices designed to fool the eye. Using sneaky techniques, you can also put on a science show that seems like magic with virtually anything at hand.

You'll learn how to make sneaky apparatuses and gimmicks for greater effect, using everyday items, to make items amazingly balance on your finger, magically levitate objects, demonstrate mind over matter, and much more.

And, you'll find simple self-working tricks that are easy to do with no preparation, including how to balance a coin on the edge of a dollar bill, create moving artwork and floating photo displays, fold origami that magically moves, and more mysterious manipulations.

Unlike simple magic tricks that are designed to fool your audience and that must be kept secret, you'll *want* to share the scientific principles and technology of your sneaky designs.

BALANCING SODA CAN

In this surprising trick, a soda can balances at a 45-degree angle on a surface or on your hand.

What's Needed
> ▶ Soda can
> ▶ Water

soda can

water

What to Do

First, pour out two-thirds of the soda from the can. See **Figure 1**. If you start with an empty can, pour water in it until it's one-third full.

Next, tilt the can on its edge so it stays in place as shown in **Figure 2**. If the can does not balance, add or remove liquid until the can stands on its edge.

Try balancing the can on a surface and even on your finger. For dramatic effect, act as if it takes lots of effort and skill to keep the can balanced.

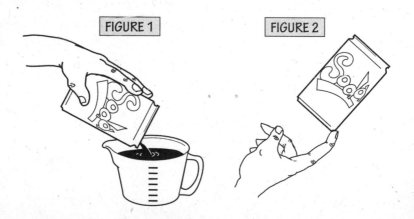

FIGURE 1   FIGURE 2

MAGICALLY JOIN TWO BOOKS

Ask a friend, especially a strong one, to take this simple challenge and watch the fun when he or she can't accomplish such a seemingly easy task.

What's Needed

▶ Two paperback books, similar in size

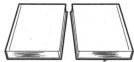

two paperback books

What to Do

First, position the books so they face each other, as shown in **Figure 1**. Then, using your thumbs, pull up the book pages and let them fall alternately, as shown in **Figure 2**. Next, when all of the pages of both books are properly joined, slowly slide the books together as closely as you can until they overlap, as shown in **Figure 3**. Press down and rub the top book cover to remove most of the air between the pages.

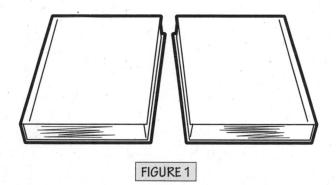

FIGURE 1

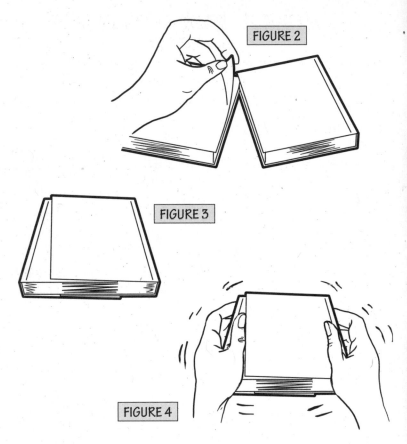

FIGURE 2

FIGURE 3

FIGURE 4

Last, ask your friend to pull the books apart for you. The person will not be able to do it. The air pressure surrounding the books and the friction make it virtually impossible to pull the books apart. See **Figure 4**.

To separate the two books, you must ruffle the pages as you slide the books apart.

MAGICAL MONEY MOTOR

You can amaze your friends by seemingly moving money with your hand without touching it. Even better, perform the trick with someone else's currency, so the individual knows it's not a trick bill.

What's Needed

- Large paper clip
- Dollar bill
- Small metallic magnet
- Plastic ring
- Tape

magnet

plastic ring

dollar bill

paper clip

tape

What to Do

Bend the paper clip into a small vertical stand, as shown in **Figure 1**. Fold and unfold a bill lengthwise, then fold and unfold the bill in half. See **Figure 2**.

FIGURE 1

FIGURE 2

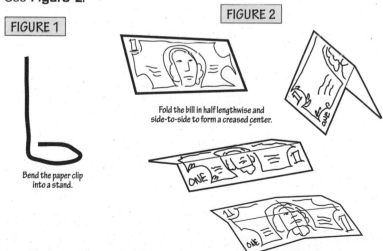

Bend the paper clip into a stand.

Fold the bill in half lengthwise and side-to-side to form a creased center.

Place the bill on top of the paper clip stand so it will balance and spin easily when turned, as shown in **Figure 3**. Next, tape the magnet on top of the toy ring. See **Figure 4**. Position the ring on your finger so the magnet is on the palm side, as shown in **Figure 5**.

Last, wave your hand around in a circular motion above the bill. It should spin because many currencies have iron particles in the ink. This is true of U.S., UK, European Union, Russian, and Brazilian paper currency. See **Figure 6**.

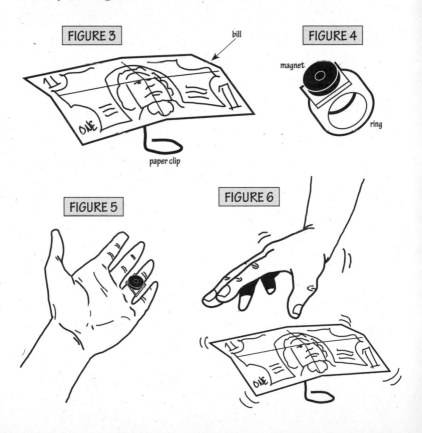

FIGURE 3

bill

paper clip

FIGURE 4

magnet

ring

FIGURE 5

FIGURE 6

STATIC ELECTRICITY TRICKS

If you rub two objects against each other, electron particles from one object will transfer to the other and become more negatively charged. The other object in turn becomes more positively charged. This transfer of charges (causing the imbalance of negative to positive charges) is termed *static electricity*.

Static electricity can transfer in sudden bursts of energy when the weather is cool and dry. You can use static electricity to perform some interesting feats of sneaky magic with common items in the home.

Salt Levitation

What's Needed
- Salt
- Cloth napkin
- Large plastic spoon or nylon comb
- Woolen sweater

salt

cloth napkin

plastic spoon

woolen sweater

What to Do
Pour salt on the napkin. Then, rub the plastic spoon back and forth against the sweater (or your hair) at least thirty times.
Lower the spoon near the salt and watch the granules magically jump to it and stick to the bottom.
See **Figure 1**.

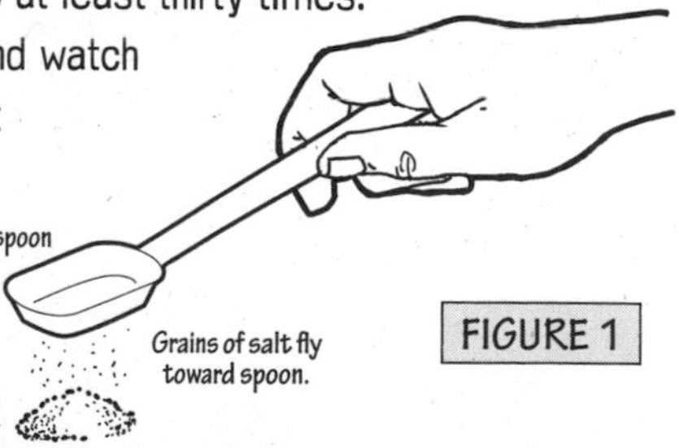

spoon

Grains of salt fly toward spoon.

salt

FIGURE 1

Water Bender

What's Needed
- Large plastic spoon or nylon comb
- Woolen sweater
- Source of running water (a faucet or hose)

plastic spoon

woolen sweater

What to Do
Rub the spoon about fifteen times against the woolen sweater. Turn on the water faucet to produce a very narrow stream of running water. Bring the spoon near the center of the stream. Watch closely and you'll see that the water will bend toward the spoon, as shown in **Figure 2**. Why does the spoon attract the stream of water? Flowing water also carries a small charge of electrical energy!

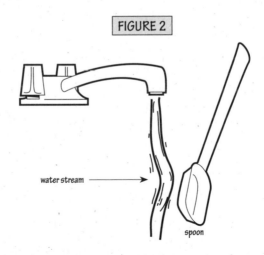

FIGURE 2

water stream →

spoon

Soda Can Magnet

What's Needed

- Large, inflated balloon
- Woolen sweater
- Empty 12-ounce aluminum can
- Table with a smooth surface

balloon

soda can

woolen sweater

What to Do

Rub the balloon against your sweater. Lay the empty aluminum can on its side on a tabletop (without a tablecloth). Slowly bring the balloon close to the can and you'll see it move. Carefully pull the balloon toward you and the can will follow in the path of the balloon. See **Figure 3**.

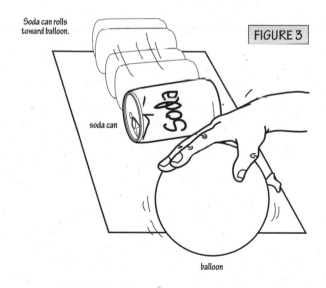

Soda can rolls toward balloon.

FIGURE 3

soda can

balloon

JUMPING TADPOLE ORIGAMI

Origami models are fun to make. But the ones that give us added pleasure are animated figures.

With just a few paper folds, you can make a cute tadpole model that jumps up in the air when you press down on it.

What's Needed
- ▶ Paper
- ▶ Scissors
- ▶ Pencil or pen

paper

scissors

pen

What to Do

Cut out a 4-inch square piece of paper. Fold the paper in half from the top and side and then unfold it. See **Figure 1**. Fold the corner sections toward the center. See **Figure 2**. Then, fold the left and right sides to the center, as shown in **Figure 3**.

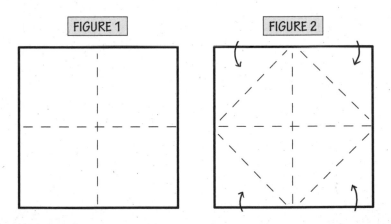

FIGURE 1

FIGURE 2

Next, fold the bottom third section up and crease it. See **Figure 4**. Fold the bottom left and right corner sections toward the center, as shown in **Figure 5**.

Last, fold the bottom section up and then half-way back down again, as shown in **Figures 6** and **7**. You can then use a pencil to draw eyes on the sneaky tadpole figure.

To make the tadpole jump, simply push down on its back, slide your finger back, and watch the little baby frog leap in the air, as shown in **Figure 8**.

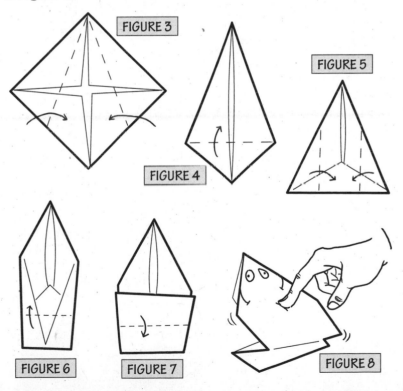

FIGURE 3

FIGURE 5

FIGURE 4

FIGURE 6

FIGURE 7

FIGURE 8

SNEAKY ANIMATED CASSETTE TAPE CREATION

Magnetism is used to record and play back tape recordings. Tiny iron particles on recording tape align themselves according to the signal placed there by the magnetic tape recording heads. On playback, the now-magnetized material in the tape moves across the tape head, which has coils of wire inside, and the signal is detected and amplified by the recorder to produce sound.

You can make some sneaky craft designs with strips of tape and animate them with a strong magnet.

What's Needed
▶ Cassette tape
▶ Strong magnet
▶ Cardboard
▶ Pencil
▶ Glue
▶ Scissors

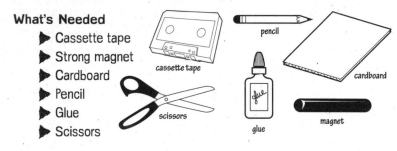

What to Do
First, draw a figure of your choosing. In **Figure 1**, an illustration of a man is shown.

FIGURE 1

Next, cut 2- to 3-inch strips of cassette tape and place glue on one end of the tape strips. Press the strips on the figure to create hair and a beard as shown in **Figure 2**. Let the sneaky design dry for 30 minutes.

FIGURE 2

Then, bring a magnet close to the drawing near the top of the figure's head. As shown in **Figure 3**, the "hair" made from the cassette tape will stand up. Try different drawings and tape arrangements to see what other animated illustrations you can create.

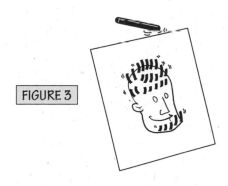

FIGURE 3

SNEAKY SODA CAN REFILL

This trick, when properly performed, will astonish onlookers, as it seems that there's no way that you can perform it without real magic. You and your onlooker will "discover" a crumpled old soda can in the trash, shake it, and turn it magically into a sealed, fully restored container that you can open and pour soda from.

What's Needed

- Black paper
- Scissors
- Unopened soda can
- Glue stick or soft glue
- Standard screwdriver

scissors

screwdriver

glue stick

black paper

soda can

What to Do

Cut a small oval piece of black paper to be the same shape as the can's opening and glue it into place on the can. See **Figure 1**.

Puncture the soda can near the top with a screwdriver to create a small hole, by carefully pushing and turning the edge on the can's surface, as shown in **Figure 2**. Pour out one-third of the soda.

Bend and crumple the can so it appears to be a discard. Position it near the top of the refuse in a trash container so it can be seen easily. See **Figure 3**.

Now, show a friend the seemingly empty can and say you can magically refill it. Pick it up casually and show the "opened" pop-top (really the black paper over a closed one), indicating that it's "empty." Next, hold the can straight out with your arm straight. With your thumb covering the small hole near the top edge, shake the can vigorously. See **Figure 4**. In about a minute, the carbon dioxide in the can will mix with the soda, causing the can to expand to normal size, as shown in **Figure 5**.

Quickly slide off the black paper covering the opening and hide it in your hand, as shown in **Figure 6**. Then, pop the top and pour out the remaining soda, and it'll seem as if the can was full of soda.

FIGURE 1

can

Stick paper over can opening area with glue stick.

FIGURE 2

Soda

Puncture small hole near top of can with screwdriver.

FIGURE 3

Soda

Bend can slightly and place on top of trash bin.

FIGURE 4

Soda

Hold thumb over hole of "empty" can and shake it.

FIGURE 5

Soda Soda Soda

Shaking can causes carbonation to expand it.

FIGURE 6

can

Soda

Quickly (and secretly) rub off black paper.

SNEAKY MONEY BALANCE TRICK

Challenge friends to fold a dollar in half and balance a quarter on its edge. They won't be able to do it. Then you'll take the same bill and the coin will stay aloft. It's very easy when you know the sneaky trick.

What's Needed

▶ Dollar bill
▶ Quarter

dollar bill   quarter

What to Do

Fold a dollar in half lengthwise, as shown in **Figure 1**. Then, fold the bill in half right to left. See **Figure 2**. Unfold the bill until the fold opens to a 45-degree angle and rest the quarter on top of it, as shown in **Figure 3**.

Now carefully pull the bill with both hands until it's nearly straight. With a little practice you can prevent the quarter from falling off, as shown in **Figure 4**. The second crease in the bill provides a second balance point, which prevents the quarter from tipping over to either side.

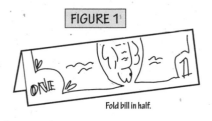

Fold bill in half.

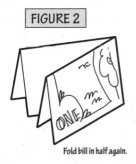

Fold bill in half again.

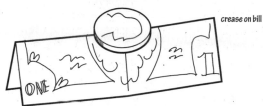

FIGURE 3

crease on bill

Unfold bill and lay quarter in middle.

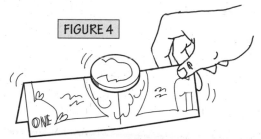

FIGURE 4

Coin will balance on bill's edge even when
pulled straight because of crease.

SNEAKY FLOATING PHOTOS

With just a few simple items you can make a floating display box for
your favorite small photographs. It will keep admirers in awe of how it
defies gravity.

What's Needed

▶ Tea box
▶ Scissors
▶ Black construction paper
▶ Glue
▶ Tape
▶ Strong magnet,
  neodymium-type preferred
▶ Paper clip
▶ Thin black thread
▶ Two (2-inch-square) photos

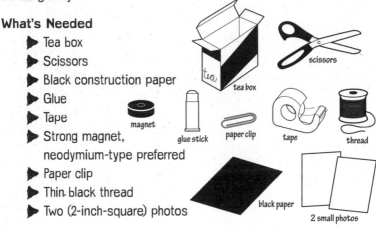

tea box

magnet

glue stick paper clip tape

scissors

thread

black paper

2 small photos

What to Do

Unfold the tea box and cut off the top and side flaps. See **Figure 1**.
Lay it flat and cut a piece of black construction paper to cover the
inside of the tea box. Glue the paper to the cardboard and let it dry.
Then cut slits in the paper to match the ones on the tea box. See
Figures 2 and **3**.

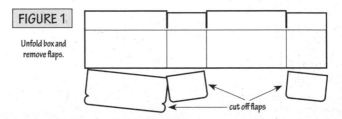

FIGURE 1

Unfold box and
remove flaps.

cut off flaps

Fold the tea box back into shape and glue it securely. The dark interior of the box will be a showcase for your photos. Tape the magnet to the inside top of the box and cover it with a small piece of black paper, as shown in **Figure 4**.

Tie a 3-inch length of thread to the end of the paper clip. Then, cover the paper clip with the two photos and tape them together, as shown in **Figure 5**.

FIGURE 2

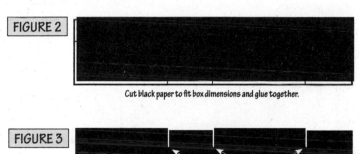

Cut black paper to fit box dimensions and glue together.

FIGURE 3

cut 3 slits in paper

Cut 3 slits in paper to match ones on tea box.

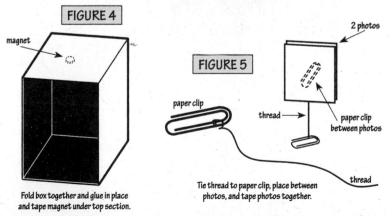

FIGURE 4

magnet

Fold box together and glue in place
and tape magnet under top section.

2 photos

FIGURE 5

paper clip

thread →

paper clip
between photos

thread

Tie thread to paper clip, place between
photos, and tape photos together.

Tape the loose end of the thread to the bottom center inside the box so the photos, when lifted to the top, have a 1/2-inch gap between them and top of the box. Lift the photos to the top and let go. Since the paper clip inside them is attracted to the magnet above, they will float in midair, as shown in **Figure 6**. If necessary, adjust the length of the thread for a proper fit. When placed on a shelf, the black thread will be virtually invisible against the black backdrop and the photos will appear to defy gravity.

FIGURE 6

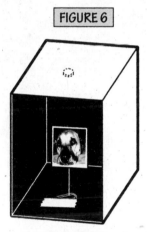

Tape thread to bottom of box and
the photo will float like magic.

DUAL FLOATING PHOTO DISPLAY

Floating photo displays are fun and eye-catching. When magnets are inserted inside two photos, they will attract each other and keep themselves suspended in midair. Here we will create a magical effect by arranging two photos to levitate with nothing above, below, or between them!

What's Needed

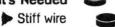

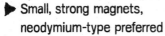

- ▶ Stiff wire
- ▶ Pliers
- ▶ Small, strong magnets,
  neodymium-type preferred
- ▶ Two small photos or illustrations
- ▶ Tape
- ▶ Nylon "invisible" thread

magnets   nylon thread   pliers   tape

two small photos   stiff wire

What to Do

In this design, you will bend a 16-inch-long piece of stiff wire into a stand. Use pliers to bend the center to a 90-degree angle. Each section of the stand, including the two vertical ends and the two center sections that form a right angle, should be 4 inches in length. The stand should appear like the one shown in **Figure 1**.

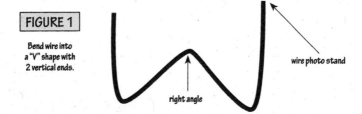

FIGURE 1

Bend wire into
a "V" shape with
2 vertical ends.

right angle

wire photo stand

Place two small magnets side by side to verify their north and south poles. They should cling to each other. Test the placement and tape the two magnets together, before going on to the next step.

Use photos printed on hard stock with an equal length of blank paper that can be folded over the back of the photo to conceal the magnets that will be placed inside. Position the magnets near the fold inside each photo display so when the photos are side-by-side, they will attract each other as shown in **Figure 2**.

Note: Make sure the magnets attract each other; turn one over until they do. Secure the magnets inside the photo displays with tape. Then tape a 6-inch length of thread to each side of the photos. See **Figure 3**.

Next, tape the end of the photograph threads to each side of the wire stand and suspend them in the air at the same height. Wrap the threads around the stand until the photos are just close enough to attract each other and levitate in midair, as shown in **Figure 4**.

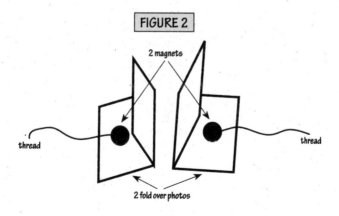

FIGURE 2

2 magnets

thread

thread

2 fold over photos

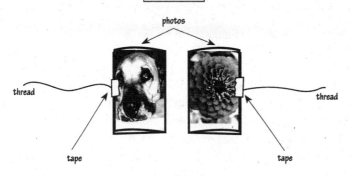

FIGURE 3

photos

thread

thread

tape

tape

FIGURE 4

SNEAKY MAGIC WAND

Every magician needs a magic wand. Sneaky magicians can make a wand from things found around the house to save resources. The wand will be put to good use when you use it to activate your cassette tape creations, floating photos, and activate devices connected to a sneaky switch.

What's Needed

- Two jumbo drinking straws
- Tape
- Two pencils
- Small, strong magnet
- Wrapping paper
- Scissors

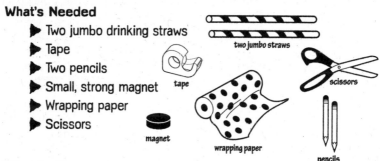

two jumbo straws

scissors

tape

magnet

wrapping paper

pencils

What to Do

Tape the two straws together end to end. Push two pencils, erasers facing out, into each end of the straws. See **Figure 1**. Place the magnet at one end of the straw and tape both ends of the straws. See **Figure 2**.

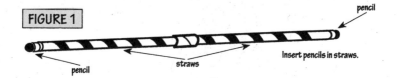

FIGURE 1

pencil

pencil

straws

Insert pencils in straws.

Tape a small piece of wrapping paper to both ends of the straw. Cut a piece of paper the same length of the straws and cover it. Tape it securely, as shown in **Figures 3** and **4**. Now you have a simple and cheap magic wand that can activate devices, float photos, and more. See **Figure 5**.

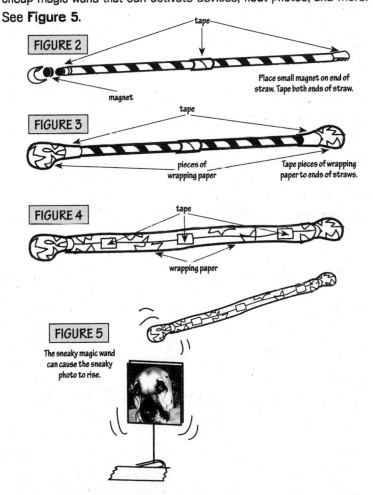

tape

FIGURE 2

Place small magnet on end of straw. Tape both ends of straw.

magnet

FIGURE 3

tape

pieces of wrapping paper

Tape pieces of wrapping paper to ends of straws.

FIGURE 4

tape

wrapping paper

FIGURE 5

The sneaky magic wand can cause the sneaky photo to rise.

SNEAKY MIND-OVER-MATTER

This is probably the best trick in the book. You'll set a small object, such as a box, down on a table, point at it with one hand, and it will magically slide over to your other hand as if you have the power of mind over matter.

What's Needed

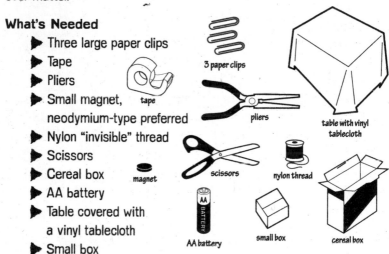

- ▶ Three large paper clips
- ▶ Tape
- ▶ Pliers
- ▶ Small magnet, neodymium-type preferred
- ▶ Nylon "invisible" thread
- ▶ Scissors
- ▶ Cereal box
- ▶ AA battery
- ▶ Table covered with a vinyl tablecloth
- ▶ Small box

3 paper clips

tape

pliers

table with vinyl tablecloth

magnet

scissors

nylon thread

AA battery

small box

cereal box

What to Do

Tape a small paper clip inside the bottom of the box. See **Figure 1**. Bend another paper clip into a half-circle, or **C** shape, so it fits around the side of the magnet. Use the pliers to bend small eyelets on both ends. Slip the magnet into the paper clip and tie two foot-long lengths of thread in the eyelets, as shown in **Figures 2** and **3**.

FIGURE 1

box

Tape paper clip to bottom of small box.

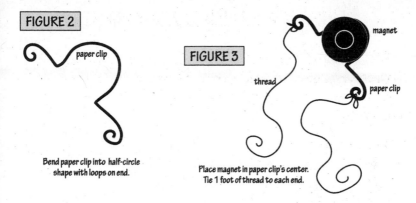

FIGURE 2

paper clip

Bend paper clip into half-circle
shape with loops on end.

FIGURE 3

magnet

thread

paper clip

Place magnet in paper clip's center.
Tie 1 foot of thread to each end.

Next, unfold the cereal box into a flat sheet and cut off the top and
bottom flaps. Fold the box in half, with a center crease and fold ridges in
it. Ridge 1 is 1/2 inch from the right side, ridge 2 is 2 inches to the left,
and ridge 3 is directly next to ridge 2. The ridges will form a path for the
magnet and thread. Apply a long piece of tape on the entire length of
the path to provide a slick surface. See **Figures 4A** through **4E**.

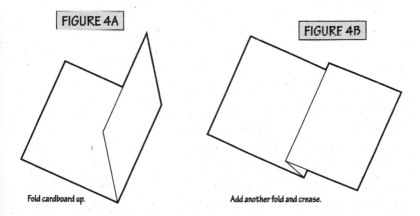

FIGURE 4A

Fold cardboard up.

FIGURE 4B

Add another fold and crease.

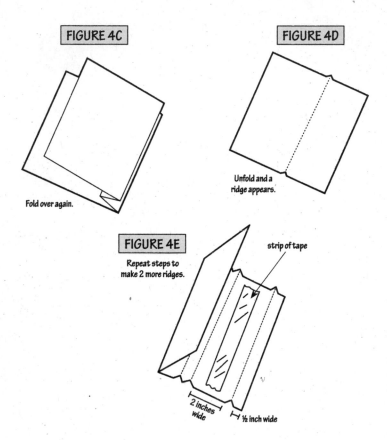

FIGURE 4C

Fold over again.

FIGURE 4D

Unfold and a ridge appears.

FIGURE 4E

Repeat steps to make 2 more ridges.

strip of tape

2 inches wide

½ inch wide

Place the magnet, and its paper clip holder and thread, in the 2-inch-wide area between ridges 1 and 2. Feed the lower thread so it flows off the front of the cardboard. Feed the upper thread back around outside the cardboard and into the path between ridges 2 and 3, as shown in **Figure 5**.

Then, tie a paper clip to the thread to the left. Tape the end of the thread on the right around the AA battery. See **Figure 6**.

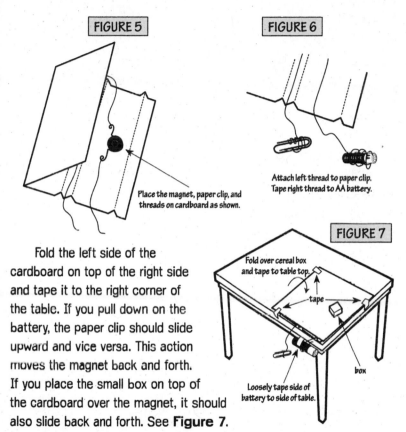

FIGURE 5

Place the magnet, paper clip, and threads on cardboard as shown.

FIGURE 6

Attach left thread to paper clip.
Tape right thread to AA battery.

FIGURE 7

Fold over cereal box
and tape to table top.

tape

box

Loosely tape side of
battery to side of table.

Fold the left side of the cardboard on top of the right side and tape it to the right corner of the table. If you pull down on the battery, the paper clip should slide upward and vice versa. This action moves the magnet back and forth. If you place the small box on top of the cardboard over the magnet, it should also slide back and forth. See **Figure 7**.

If necessary, roll the thread around the battery and tape it securely. Pull all the way on the paper clip and the battery should pull up to the edge of the cardboard. Drop the battery and the paper clip will slide upward.

Place the box on top of the tablecloth where the magnet is positioned and it will "stick" in place. Roll up a small ball of tape and stick to the side of the battery. Press the battery to the side of the table. See **Figure 8**.

Now for the display of mind over matter. Point at the box with your left hand, while your right hand is resting on the table. Secretly push on the battery under the tablecloth with your little finger until it works free. See **Figure 9**.

When the battery is loosened from the table, it will start to fall. You can easily control the battery's descent by using your finger to adjust the pressure against the thread. See **Figure 10**. If the box does not move along with the magnet, tape another paper clip to it.

As the battery falls, it pulls the magnet along and the box follows it. You can grab the box with your right hand and pick it up with a single motion like magic, as shown in **Figure 11**.

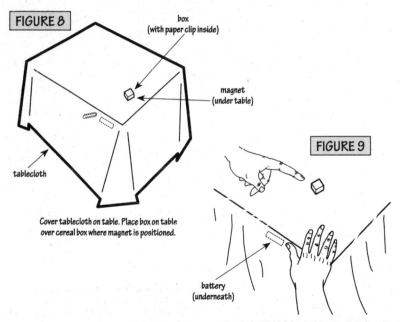

FIGURE 8

box
(with paper clip inside)

magnet
(under table)

tablecloth

Cover tablecloth on table. Place box on table
over cereal box where magnet is positioned.

FIGURE 9

battery
(underneath)

Point at box with left hand, while secretly pushing
battery from side of table under the tablecloth.

FIGURE 10

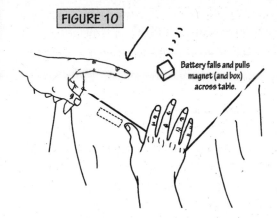

Battery falls and pulls magnet (and box) across table.

Once the battery is pushed loose, it falls and pulls the magnet toward the edge of table. The paper clip in the box follows the magnet and the box magically slides into right hand.

FIGURE 11

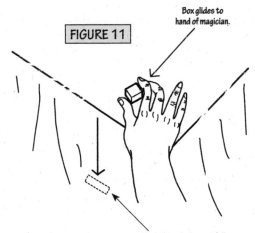

Box glides to hand of magician.

Paper clip rises under tablecloth as the battery falls. To reset for another time, pull paper clip downward and tape battery (loosely) to table's side.

SNEAKY TRIVIA

This collection of sneaky facts and science trivia is selected not to just puzzle your friends but to make you think a little differently about the things around you. You'll relish and share the amusing pictures showing fun facts and details.

Some of them seem obvious, such as: Which state is farthest north, east, and west? Guess what? It's not Washington, California, Florida, or Maine. In fact, it's a single state!

You'll learn which animal Is the only onc whose testimony is used in court, the true distance between Russia and America, how to tell the final fate of a person who is shown riding a horse in a statue, how many drops of water can fit on a penny, and much more.

For your amusement and others, this collection of Sneaky Trivia will keep you guessing until the end.

WATER ON A PENNY

You can place 40 drops of water on the head of a penny.

Water has two hydrogen atoms and one oxygen atom, hence, H_2O. The extraordinary stickiness of water is due to the two hydrogen atoms, which are arranged on one side of the molecule and are attracted to the oxygen atoms of other nearby water molecules in a state known as *hydrogen bonding*. If the molecules of a liquid did not attract one another, then the constant thermal agitation of the molecules would cause the liquid to instantly boil or evaporate.

Surface tension plays an important role in the way liquids behave. If you fill a glass with water, you will be able to add water above the rim of the glass, because of surface tension. Hydrogen bonds between water molecules form a "skin" on the surface of the water. When you carefully place additional water droplets on the penny's surface, the molecules form a bond that causes the water to form a blob. Rather than spilling off the edge of the penny, the blob of water clings together and grows vertically until the force of gravity eventually causes it to break apart. See **Figure 1**.

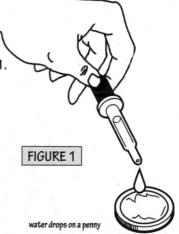

FIGURE 1

water drops on a penny

THE HEIGHT OF THE EIFFEL TOWER

The height of the Eiffel Tower varies by as much as 6 inches.

Because the tower is made of iron and metal, it can expand and contract without damage depending on the temperature: all objects expand when hot and shrink when cool. Because of this, the Eiffel Tower is taller in the summer. This is why some sidewalks and pavement can crumble in extreme summer heat. The expanding pavement has nowhere to expand to and cracks under pressure. (Separation lines are placed on some city street sidewalks to allow for some heat expansion and to prevent damage.) See **Figure 2**.

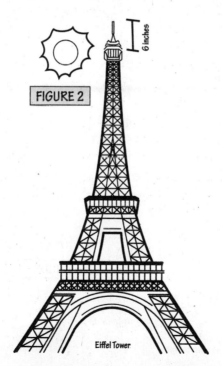

FIGURE 2

6 inches

Eiffel Tower

BODY MEASUREMENTS

| FIGURE 3A | The measurement from your wrist to your elbow is the same as the length of your foot. Also, your thumb is the same length as your nose. See **Figures 3A** and **B**. |

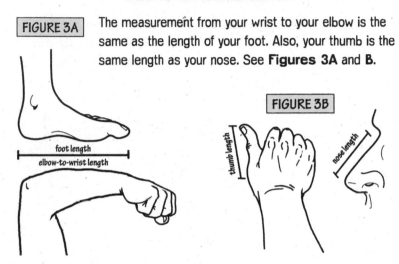

foot length

elbow-to-wrist length

FIGURE 3B

thumb length

nose length

STATUES

A statue of a soldier on horseback indicates how he died.

If the horse has both front legs in the air, the person died in battle; if the horse has one front leg in the air, the person died as a result of wounds received in battle; if the horse has all four legs on the ground, the person died of natural causes. See **Figure 4**.

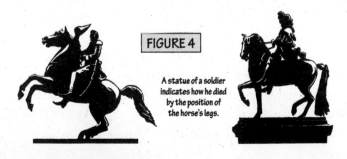

FIGURE 4

A statue of a soldier indicates how he died by the position of the horse's legs.

RUSSIA AND AMERICA

Russia and America are less than 2.4 miles apart at the Bering Strait, which joins the Bering Sea to the Chukchi Sea on either side. On one side, Alaska's Little Diomede Island is 2.4 miles from Russia's Big Diomede Island. See **Figure 5**.

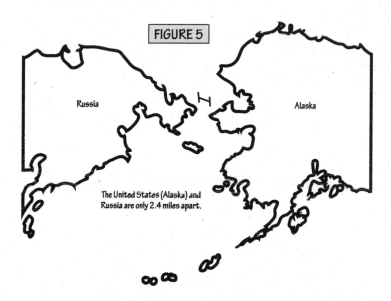

FIGURE 5

Russia

Alaska

The United States (Alaska) and Russia are only 2.4 miles apart.

ALASKA

Alaska is the northernmost state of the United States. It is also the westernmost and easternmost state, the latter because the Aleutian Islands extend past 180 degrees longitude, the international dateline that distinguishes east from west. (In case you were wondering, Hawaii is the southernmost state in the USA.)

Alaska is also the largest U.S. state: Rhode Island could fit into Alaska 425 times. At 365 million acres, Alaska is so large that if you crossed 1 million acres per day, it would take you a year to cover all of the state.

This state is also so big it has its own time zone: Alaska Standard Time (AST)—in fact, the international timeline had to be bent to keep all of the state in the same day! Alaska standard time is one hour behind Pacific Standard Time. See **Figure 6**.

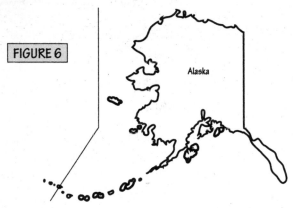

FIGURE 6

Alaska

Extra Sneaky Fact

Alaska is so big it can fit twenty-one other states inside it all at once:

Arkansas	Maine	Pennsylvania
Alabama	Maryland	Rhode Island
Connecticut	Massachusetts	South Carolina
Delaware	Mississippi	Tennessee
Hawaii	New Hampshire	Vermont
Indiana	New Jersey	Virginia
Kentucky	Ohio	West Virginia
Louisiana		

A BOEING 747'S WINGSPAN

A Boeing 747's wingspan—211 feet—is almost twice as long as the entire Wright Brothers' first flight (112 feet). See **Figure 7.**

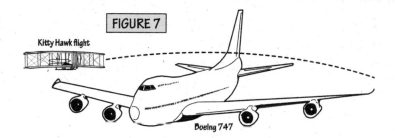

FIGURE 7

Kitty Hawk flight

Boeing 747

BAMBOO

Bamboo can grow up to 35 inches in a single day.

Botanically speaking a member of the grass family, it is the fastest-growing woody plant in the world, because of its rhizome root system. See **Figure 8.**

FIGURE 8

Bamboo can grow up to 3 feet in 1 day.

WATERMELONS IN JAPAN

Watermelons are grown square in Japan to save space when stacked.

The square watermelons are created by inserting them into cube-shaped glass containers while they are still growing. Square watermelons will fit more easily on shelves and therefore save refrigerator space. Each melon sells for 10,000 yen, equivalent to about $83, almost ten times the price of a normal U.S. watermelon. See **Figure 9**.

FIGURE 9

square watermelon

THE BLOODHOUND

The bloodhound is the only animal whose testimony is permissible in court.

The bloodhound's ability to read terrain with its nose is due primarily to a large, ultrasensitive set of scent membranes that allow the dog to distinguish smells at least a thousand times better than humans can. Whereas our olfactory center is about the size of a postage stamp, a dog's can be as large as a handkerchief. Using an odor "image" as a reference, the bloodhound is able to locate a subject's trail, which is made up of a chemical cocktail of scents from the person's breath, sweat vapor, and skin, among other things.

FIGURE 10

bloodhound in court

Once the bloodhound identifies the trail, it will not divert its attention despite being assailed by a multitude of other odors. Only when the dog finds the source of the scent or reaches the end of the trail will it relent. So potent is their drive to track, bloodhounds have been known to stick to a trail for more than 130 miles! See **Figure** 10.

MOSQUITOES

Mosquitoes have caused more human deaths than has any other creature.

According to the Centers for Disease Control and Prevention, over 1 million people die from mosquito-borne diseases every year. Not only can mosquitoes carry germs that afflict humans, they also transmit several diseases and parasites to which dogs and horses are very susceptible. See **Figure** 11.

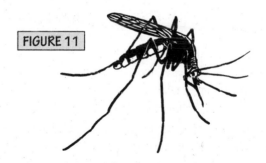

FIGURE 11

THE HORNED LIZARD

To scare off enemies, the horned lizard squirts blood from its eyelids.

It uses a series of thin-walled, blood-filled spaces called *sinuses*, found within its eye sockets. When the lizard rapidly increases the blood pressure within these sinuses, it causes the sinus walls to rupture suddenly. The blood is then forced out in jetlike squirts of crimson droplets. Sometimes the force with which the lizard squirts this eye-ejected blood is so powerful that it can send sprays shooting up to distances of 4 feet. This bizarre squirting can be repeated several times if necessary, which is usually sufficient to frighten off any predator. Also, the squirted blood contains a distasteful chemical, which would act as an additional deterrent to potential predators. See **Figure 12**.

FIGURE 12

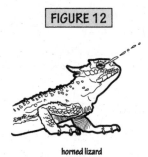

horned lizard

SHARK EMBRYOS

The embryos of tiger sharks fight in the womb. Only the survivor is born.

The sand tiger shark produces as many as twenty-five thousand pea-sized eggs in a lifetime. Periodically, fifteen or twenty eggs pass from the ovary into each oviduct, where they are fertilized and packaged inside an avocado-shaped egg case, where the shark embryos begin to develop.

It is then, even though they are tiny, that their struggle for survival begins. For most of them it does not last long. The embryos begin eating one another until only one, the fiercest and fittest, remains. It does not starve, for soon a new egg case comes down the oviduct and is promptly eaten.

After a yearlong succession of egg-case deliveries, the baby shark in each of the left and right oviducts is 40 inches long—close to half the length of its 8-foot mother. See **Figure 13**.

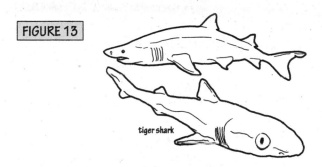

FIGURE 13

tiger shark

HUMMINGBIRDS

The hummingbird is the only bird that can fly backward.

A hummingbird can rotate each of its wings in a circle, allowing it to be the only bird that can fly forward, backward, up, down, or sideways, or hover in the air. To hover, the hummingbird moves its wings forward and backward in a repeated figure eight, much like the arms of a swimmer treading water.

Hummingbirds can move instantaneously in any direction, start from their perch at full speed, and don't necessarily slow up to land. They can even fly short distances upside down, a rollover trick they employ when being attacked by another bird. See **Figure 14**.

FIGURE 14

The hummingbird is the only bird that can fly backward.

THE MAYFLY

An adult mayfly lives for just one day.

The lifespan of an adult mayfly can vary from just thirty minutes to one day, depending on the species. See **Figure 15**.

FIGURE 15

The mayfly lives for one day.

MAGNETIC BIRDS

Many birds and animals and some insects and fish have tiny magnetic particle-carrying bacteria in their heads that act like a compass.

Studying bacteria that lived in the muds of ponds and marshes, Salvatore Bellini found little rodlike bacteria that all swam together in one direction—north. More study showed that each little bacterium had a chain of dense particles inside. And the particles proved to be *magnetite*, a magnetic form of iron oxide.

FIGURE 16

Some birds have magnetic particles in their heads that aid in their sense of direction

The bacteria had made themselves into little magnets that could line up with the Earth's big magnet. The big news was that a living thing, even a simple little bacterium, can make magnetite. That led to a search to see whether animals might have it. See **Figure 16**.

WORLD CITY LATITUDE LIST

DEGREES/MINUTES		CITY	COUNTRY
64	04 N	Reykjavik	Iceland
60	10 N	Helsinki	Finland
59	57 N	Oslo	Norway
59	56 N	Saint Petersburg	Russia
59	17 N	Stockholm	Sweden
57	09 N	Aberdeen	Scotland
55	55 N	Edinburgh	Scotland
55	45 N	Moscow	Russia
55	40 N	Copenhagen	Denmark
54	58 N	Newcastle-on-Tyne	England
54	37 N	Belfast	Northern Ireland
53	33 N	Hamburg	Germany
53	20 N	Dublin	Ireland
52	30 N	Berlin	Germany
52	25 N	Birmingham	England
52	22 N	Amsterdam	Netherlands
52	14 N	Warsaw	Poland
51	32 N	London	England
50	52 N	Brussels	Belgium
50	05 N	Prague	Czech Republic
48	48 N	Paris	France
48	14 N	Vienna	Austria
47	30 N	Budapest	Hungary
47	21 N	Zurich	Switzerland
45	26 N	Venice	Italy
45	30 N	Montreal	Canada
44	52 N	Belgrade	Yugoslavia

44	25 N	Bucharest	Romania
43	40 N	Toronto	Canada
42	40 N	Sofia	Bulgaria
41	54 N	Rome	Italy
41	23 N	Barcelona	Spain
40	26 N	Madrid	Spain
40	47 N	New York	USA
39	55 N	Ankara	Turkey
39	55 N	Beijing	China
38	44 N	Lisbon	Portugal
38	53 N	Washington, D.C.	USA
37	58 N	Athens	Greece
36	50 N	Algiers	Algeria
35	45 N	Teheran	Iran
35	40 N	Tokyo	Japan
34	03 N	Los Angeles	USA
32	57 N	Tripoli	Libya
31	10 N	Shanghai	China
30	02 N	Cairo	Egypt
23	08 N	Havana	Cuba
22	34 N	Calcutta	India
21	29 N	Mecca	Saudi Arabia
19	26 N	Mexico City	Mexico
19	00 N	Mumbai	India
17	59 N	Kingston	Jamaica
14	40 N	Dakar	Senegal
14	35 N	Manila	Philippines
13	45 N	Bangkok	Thailand
10	28 N	Caracas	Venezuela

8	58 N	Panama City	Panama
4	32 N	Bogotá	Colombia
1	14 N	Singapore	Singapore
0		Equator	
2	10 S	Guayaquil	Ecuador
4	18 S	Kinshasa	Congo
6	16 S	Jakarta	Indonesia
9	25 S	Port Moresby	Papua New Guinea
12	56 S	Salvador	Brazil
16	27 S	La Paz	Bolivia
18	50 S	Antananarivo	Madagascar
20	10 S	Iquique	Chile
22	57 S	Rio de Janeiro	Brazil
23	31 S	Sao São Paulo	Brazil
26	12 S	Johannesburg	South Africa
27	29 S	Brisbane	Australia
31	57 S	Perth	Australia
33	28 S	Santiago	Chile
33	55 S	Cape Town	South Africa
34	00 S	Sydney	Australia
34	35 S	Buenos Aires	Argentina
34	53 S	Montevideo	Uruguay
34	55 S	Adelaide	Australia
36	52 S	Auckland	New Zealand
37	47 S	Melbourne	Australia
41	17 S	Wellington	New Zealand
42	52 S	Hobart	Australia

POPULAR U.S. CITY LATITUDE LIST

CITIES	LATITUDE	LONGITUDE	TIME
	(decimal and degrees/minutes)		

(**Example:** Chicago, Ill. 41° 50' N 87° 37' W at 11:00 A.M.
Notice that all of the cities below are degrees north and degrees west
because they are all in North America.)

Anchorage, Alaska	61° 13' N	149° 54' W	8:00 A.M.
Atlanta, Ga.	33° 45' N	84° 23' W	12:00 noon
Baltimore, Md.	39° 18' N	76° 38' W	12:00 noon
Boise, Idaho	43° 36' N	116° 13' W	10:00 A.M.
Boston, Mass.	42° 21' N	71° 5' W	12:00 noon
Buffalo, N.Y.	42° 55' N	78° 50' W	12:00 noon
Chicago, Ill.	41° 50' N	87° 37' W	11:00 A.M.
Cincinnati, Ohio	39° 8' N	84° 30' W	12:00 noon
Dallas, Tex.	32° 46' N	96° 46' W	11:00 A.M.
Denver, Colo.	39° 45' N	105° 0' W	10:00 A.M.
Fargo, N.Dak.	46° 52' N	96° 48' W	11:00 A.M.
Honolulu, Hawaii	21° 18' N	157° 50' W	7:00 A.M.
Kansas City, Mo.	39° 6' N	94° 35' W	11:00 A.M.
Las Vegas, Nev.	36° 10' N	115° 12' W	9:00 A.M.
Los Angeles, Calif.	34° 3' N	118° 15' W	9:00 A.M.
Memphis, Tenn.	35° 9' N	90° 3' W	11:00 A.M.
Miami, Fla.	25° 46' N	80° 12' W	12:00 noon
Milwaukee, Wis.	43° 2' N	87° 55' W	11:00 A.M.
Minneapolis, Minn.	44° 59' N	93° 14' W	11:00 A.M.
New Orleans, La.	29° 57' N	90° 4' W	11:00 A.M.
New York, N.Y.	40° 47' N	73° 58' W	12:00 noon
Philadelphia, Pa.	39° 57' N	75° 10' W	12:00 noon
Pittsburgh, Pa.	40° 27' N	79° 57' W	12:00 noon
St. Louis, Mo.	38° 35' N	90° 12' W	11:00 A.M.

Salt Lake City, Utah	40° 46' N	111° 54' W	10:00 A.M.
San Francisco, Calif.	37° 47' N	122° 26' W	9:00 A.M.
Washington, D.C.	38° 53' N	77° 02' W	12:00 noon
Winnipeg, MB, Canada	49° 54' N	97° 7' W	11:00 A.M.